Test Pilot at War

A

H. A. TAYLOR

TEST PILOT AT WAR

LONDON

IAN ALLAN

First Published 1970

SBN 7110 0138 3

Published by Ian Allan Ltd., Shepperton, Surrey, and printed in the United Kingdom by Balding & Mansell Ltd., London and Wisbech.

Contents

The test-flying and ferrying reminiscences and impressions of World War Two which make up this book are rearranged and suitably edited versions of a series of articles published between 1960 and 1968 in Air BP, *the journal of the International Aviation Service of the BP Group. The author wishes to thank this organization and the consultant editor of* Air BP, *John W. R. Taylor, for permission to publish these articles in book form.*

Prologue

It was a bright, cool afternoon during the early summer of 1943. A weather system, which had been a full-grown front over the open Atlantic and the Irish Sea, had lost its defined characteristics as it moved south-eastwards over the Welsh hills and was now no more than a series of cumulus clouds sailing over the fields of Shropshire.

In the test pilots' hut at an RAF maintenance unit the telephone rang and was answered by one of the three waiting pilots. From a battered armchair in the opposite corner of the room I could hear the voice at the other end of the line: "Number Two Site here; 295 is ready for test."

That was a flight for which I had put myself down in the authorisation book. I went over to check it. Yes, there it was: "Beau VI; BT295; Flt Lt Taylor. Air test and air firing".

This Bristol Beaufighter had already been flown twice and there was only one minor instrument snag to be cleared. So I could fly it straight up to that area of the Irish Sea off the coast of North Wales into which, to make sure that all the cannons worked, we periodically and somewhat wastefully fired a stipulated number of ball, or non-explosive rounds. In short, the outing was likely to be a restful one during which it might be possible, for once, to enjoy flying for its own sake.

At Number Two Site the Beaufighter stood waiting, in all its ugly, functional dignity, on an area of muddy grass beside the perimeter track. Its two Bristol Hercules radial engines, disproportionate in their size and hiding the pilot's cockpit between them, were "tick-tocking" quietly as they cooled down after the engine fitter's check run-ups. I signed the familiar Form 700 and climbed awkwardly up the trapdoor steps into the crowded seat where my parachute was already waiting. With a rigger's help I sorted out, adjusted and secured the parachute and Sutton harnesses and then waited for the sharp metallic clang which meant that the trapdoor had been closed and that the rigger was clear.

At a thumbs-up sign from the ground crew I began to go mechanically and quite automatically through the engine-starting drill. In their turn the two engines burst lumpily and smokily into life as the starter and booster-coil buttons were pressed. After a few moments of warm-up, with the oil temperatures rising, the oil pressures dropping to normal and the revolution-counter needles swinging idly between a thousand and thirteen hundred, it was time to do the final engine runs and to check manifold pressures, magnetos and constant-speed propeller operation. All this had to be done quickly and with a perfunctory air so as not to upset the feelings of the engine fitter, who had already completed the same checks to his own satisfaction.

Then the chocks were waved away and, with an inch or two of extra throttle, we rolled off the grass and on to the perimeter track with the brakes wailing thinly as their control-column trigger was squeezed. The Beaufighter jerked and squealed its way down-wind along the perimeter track, enveloped in the brightly blue wreaths of smoke from the exhaust stacks.

Before turning on to the runway in use I went through the long-established pre-take-off drill. Hydraulics "on"; three trimming controls set to zero, zero and one inch forward; mixture rich; propeller controls fully forward for maximum revolutions; inner fuel tanks selected; dive-brakes in; flaps up; and engine-cooling gills closed and then opened a trifle. All set. On the runway each engine was cleared in turn with a burst of throttle and the brakes were released just as soon as a steady green light shone from the controller's chequered box.

A touch or two of brake to get the outfit properly lined up as it started to move and then the throttles were inched open individually to keep the Beaufighter straight during the first critical part of its take-off run. After a second or two the throttles were taken through the "rated" gate to maximum power and the whole thing came to life in a burst of acceleration. At one moment it had been an ungainly, deviating monster; at the next it was plunging like a railed locomotive down the short runway.

The Beaufighter floated off the ground; undercarriage "up" was selected and the nose yawed one way and then the other as the legs came up one after the other.

There was a sense of peace and almost of quiet with any aeroplane of this earlier era as boost and revolutions were brought back to climb power after take-off, with the undercarriage snugly retracted and the airspeed-indicator needle creeping round the dial.

Sun and shade flicked by with a kaleidoscopic effect and presently there was a series of short, sharp bumps as the Beaufighter climbed through the first of the broken cumulus, with flashes of brilliant light and colour alternating with seconds of gloom when swirling vapour covered the flat, bullet-proof screen with streaks of water. Then, finally, the apocalypse – as the whole, wide vault of heaven leaped into view.

Everything seemed to be twice as brightly coloured and twice as real as life. A rivulet of oil along the port engine cowling was dancing in microscopically wrinkled, shiny brown patterns as it was lashed by the invisible slipstream. The Beaufighter seemed to be poised stationary over a cloud landscape. Only an insistent and irritating rattle from a bit of loose fabric on the centre-section broke into the overall quiet of steady, purposeful, yet unheard clamour.

Away out in front now was the coastline, with the mottled silver sea beyond it. Time to get back to work and to prepare for the cannon-firing run. There were no clouds over the sea and the test area seemed to be clear of shipping. But we had to come down quite a bit from our lordly 10,000ft before it would be safe to start a run-in.

This would be as good an opportunity as any for checking the dive-brakes by using them to slow down the Beaufighter before it hit the last of the clouds on the way down to 4,000ft. Even so, as these ragged remnants whisked by, the Beaufighter felt as if it were hitting a series of solid objects.

A final circuit was made over the firing area to be quite sure that there were

no innocent boats about, and then the Beaufighter was lined up for the exhilarating little exercise–just about the only one which offered some sort of genuine excuse for the deadly RAF crime of low flying.

Twenty-six hundred revolutions and rated boost were set up; the elevator trim was gently wound forward; the cannons were already cocked and the firing button was set to "fire". Most of the outside world was now hidden behind the fiery pink design of the gunsight, but out of the corner of an eye I could see the rate-of-climb needle over on the wrong side of its dial and the hands of the altimeter going round in reverse like those of a clock with a broken escapement. A bit more forward trim was wound on. Now was the moment to press that little button.

A rough uneven hammering shook the aircraft and acrid fumes of burnt cordite filled the cockpit. It was difficult to look through smarting and watering eyes, but, flying down-sun as we were, the shells could actually be seen as they plunged away like silver wraiths, each surrounded by an aura of shimmering smoke. A few seconds later they were making fountains of spray in the sea.

This sea was no longer a mottled plate of silver. It was real and mountainous and the waves were white-capped. Time to start pulling out for another pass.

A little nose-up trim was wound on to help the pull-out and as the horizon dropped away there was the familiar impression of darkness and dreamy heaviness with increased *g* forces. The directional gyro "toppled" and its drum spun wildly as we pulled up in a climbing turn.

A minute later, after the last of the four guns had hammered out its remaining rounds–and after the guiltless opportunity had been taken of beating up an empty beach–the Beaufighter was allowed to make its own way up to a respectable cruising altitude. As the boost and revolutions were brought back Snowdonia was already drifting away and behind to the right.

As I sat relaxed in that crowded little cockpit–idly scanning the instruments or the fields of Cheshire (each with its little "dewpond") and listening absent-mindedly to the familiar thumping sing-song note of the two Hercules engines–my thoughts went back over the years.

How on earth had I got myself into this ludicrous situation–in sole and fairly competent control of a mighty little flying engine of war? I was, after all, no more than a club-trained amateur–and near middle-aged at that–wearing a uniform for which I had suffered no square-bashing and wings which had been won by default, surrounded by a complicated airborne contraption which I had certainly never been taught to fly.

CHAPTER ONE

The Guinea Pigs

*

ATA

EVEN THOUGH I had known for a day or two that something of the kind was brewing, the words of the telegram came as something of a shock – just as they did for the other pilots, mostly club-trained amateurs, who formed the small nucleus of the Air Transport Auxiliary in September 1939. The telegram said that I was to report to the RAF Central Flying School at Upavon in two days' time for "conversion" to military aircraft with a view to possible ferrying duties – or words to that effect.

The mild shock was caused by the fact that the flying of military aircraft was not at all the work for which this organisation had been initially designed. The ATA had been formed to make wartime use of selected amateur pilots who were too old for, or otherwise unacceptable to, the powers who then controlled the Royal Air Force. For a year or more two or three enthusiasts had been badgering the Air Ministry and other authorities to find some way of using this available experience which would otherwise have been wasted. At last, in mid-August 1939, permission was given for suitable pilots to be selected and enrolled. The idea was that they would fly small aircraft – up to, say, DH89 Rapide size – and provide necessary additional pilot strength for the professional air transport services.

Hence the name of the organisation which, as things turned out, was to be less than appropriate to the work which it eventually did. Nearly everybody then thought that the war, when it inevitably started, would involve the more or less immediate and wholesale bombing of all main centres of industry and population. The ATA was expected, in some vague way, to help to maintain communications and, perhaps, to ferry people and medical supplies about the country.

Within a few days of the declaration of war on September 3 the little group had been checked out in a Tiger Moth for their basic flying abilities or otherwise. The two dozen or so who passed the tests were duly signed on as semi-professional civil pilots with British Airways (then about to be merged with Imperial Airways to become BOAC) and provided with a featureless navy-blue uniform and an RFC/RAF-type forage cap – which never felt that it could possibly stay on. By this time the whole of British air transport, as well as units from some foreign airlines, had moved complete from London's Croydon and Heston airports to Bristol, where (in two rooms of a commandeered hotel in Clifton) we signed our contracts.

At the time, all this seemed to be very satisfactory and even exciting. Some of us had, after all, tried hard but unsuccessfully to find a way of remaining airborne with the RAF. The Air Ministry of that time was, however, interested only in the young and in the ex-Service Reserve and the Volunteer Reserve pilots – though glad enough to consider others for non-flying duties. Only the Royal Navy was bright enough to realise that amateur pilots could be useful and was, by mid-September, enrolling them with the rank of sub-lieutenant in the RN Volunteer Reserve of the Fleet Air Arm – where some of them duly grew beards and became unrecognisable.

A change of Air Ministry policy was to come fairly soon and I, for one, succeeded – after an initial failure and a year's service with the ATA – in getting past the medical and selection boards and into that anonymously named general duties branch of the RAF. But in September 1939 we of the ATA were grateful enough to be allowed to fly in any capacity during a war which many of us already felt was going to last for a long, long time.

The bombs did not drop and the only aircraft to be shot down were one or two of our own Blenheims over the Thames estuary – mistaken, by trigger-nervous gunners or fighter pilots, for approaching Heinkels. Neither the ATA, not, for that matter, the airline professionals, were required to "maintain communications". But more pilots *were* needed to ferry military aeroplanes from factories to RAF maintenance units, and, later, from factories and MUs to operational units. In those days there were only two pools of ferry pilots – the RAF's Number 1 Ferry Pilots' Pool based at Hucknall, Nottingham, and No. 2 FPP at Filton, Bristol. Operational pilots were still, for the most part, collecting their own aircraft as they were needed – but there was a rapidly increasing trade between the so-called shadow factories and the MUs where aircraft were equipped, armed and stored. It was obvious, too, that, when the war eventually ceased to be "phoney", the squadron pilots would be much too busy to collect their own aeroplanes. The strength of the two RAF ferry pools would clearly need to be increased and more pools formed.

So, within fewer than 20 days of the start of the war, this most remarkably uninhibited decision was made – that amateurs should be given a chance to show whether they could learn to handle these military monsters and to ferry them. In three months or so the authorities were to make an even more remarkable decision – to approve the formation of an initially small ferry pool composed exclusively of experienced women pilots.

The six days which I spent at the RAF's Central Flying School in September 1939 were the most satisfying of any that I can remember. This was not just because I was doing something which I had always wanted to do, nor even because of the challenge which the brief conversion course presented. I had never been much attracted by "challenges" and, after all, the conversion of a few ATA guinea pigs was only a tiny part of the work that was going on at that school during the first month of the second of the wars to end all wars. No, it was the atmosphere of the place.

Here on the Wiltshire Downs were the instructors and ex-instructors who had been recalled or had volunteered for work with the RAF. Many were near-middle-aged men whose last instructional experience in the Service might have been in the back seats of Avro 504Ns (or even 504Ks) and whose most recent instructing experience, at the best, might have been in the front seats of club Gipsy or Tiger Moths. Now they were being put through the mill of the new techniques by the instructors of instructors. Not only had they to learn to fly a complicated advanced trainer, the North American Harvard, but they had to learn to instruct pupils in it, both by day and by night, using a back seat which did not, by any means, provide the best view of the world at large. There were inevitable accidents, mostly mild, and the nights were made hideous, for those who were off duty, by the scream of the Harvards–our own and those from nearby Netheravon–but the whole thing was going on against a background of enthusiasm which was suffered with admirable detachment by the RAF regulars.

We, the first pilots of the ATA, were no more than interlopers in this scene of cheerful effort and hard day and night routines, but the instructors gave us all the skill and attention they could spare and were surprisingly tolerant of our civilian background and training. It had been made clear to us (and to them) that there could be no question of our being *trained* to fly these lethal, complicated and powerful contraptions. If we did not succeed in being "converted" in a reasonably short period of time in the air we would be politely sent home. In fact, some among the early guinea pigs failed to make the grade with the Harvard–but the majority succeeded, and the first four to be tried out went on to do successful battle with a twin-engined Bristol Blenheim bomber. They and half a dozen others in the initial groups were passed out as fit to fly any aeroplane on the ferry pools' books–a result which was certainly more surprising to them than to their instructors.

Fourth among the guinea pigs, I managed in my turn to get through to the Blenheim. My logbook says (impossible though it seems now) that I was let loose alone in the Harvard after a little more than an hour's dual and that I flew it thereafter–without causing any apparent damage–for another hour or so. I was then pushed off, alone except for human ballast, in a single-engined Fairey Battle bomber and flew it without causing any obvious catastrophe. As a prospective soloist with the Blenheim I was probably a very near miss. My logbook says that I was given more than two hours' instruction before anyone dared to let me go off in it alone. The solo Blenheim flying was the greatest experience of all–but more about that later.

My logbook also tells me that my total hourage at this time was 370, of which only about 240 were genuine solo hours. The situation had no doubt been saved for me by the fact that this miserable hourage had been clocked up since 1929 in a lot of different light aeroplanes. I had also completed an instrument-flying course, gone through part of an instructor's course and flown as supernumerary second pilot in several types of then-advanced transport aircraft.

A fortnight later, in mid-October, 1939, two dozen ATA pilots–some "converted", but others not–were sent on attachment to the two RAF ferry pools. For the ATA the war had really started. □

CHAPTER TWO

Aerobatics – 40 minutes

*

HARVARD

"Now," said the instructor, "let's try a loop. Keep the wings level and don't try to hurry it over the top." This was my second conversion period in the Harvard at the RAF's Central Flying School, Upavon. During the first, the day before, I had managed to cope, after a fashion, with the complicated cockpit drill and had made half-a-dozen not too desperately untidy circuits and landings.

On this second outing – designed, I thought, to demonstrate some of the characteristics of "hot" aeroplanes with highish wing loadings – we had done a few circuits and bumps. Then I'd been talked through spins and recoveries from spins. "Back . . . back . . . away she goes. That's right: hard opposite rudder. No, don't ease the stick forward yet; stop the rotation first." The spins had been rather horrible affairs by earlier Moth, Avro Cadet and Tiger Moth standards. The Harvard had entered the spin with a strange kind of violence and had seemed to me to have rotated with a wildly irregular motion.

A loop, I thought, should be a fairly straightforward affair by comparison. You just had to hold enough backward pressure on the stick to keep your weight comfortably in the seat and watch the wing tips for yaw until the horizon re-appeared over the nose. It seemed to be a long time before the horizon appeared and by then I was beginning to float off the seat in a Sutton harness which was not really tight enough. A little more backward pressure on the stick would, I thought, keep me more firmly in the seat and take the Harvard neatly over the top of the loop.

But the Harvard had ideas of its own. It dropped a wing sharply in an incipient spin, which was stopped with instinctively applied opposite rudder. None of us – Harvard, instructor, or pseudo-pupil – lost too much face. The manoeuvre ended as a sort of half-hearted flick roll. If anyone had been watching us at 8,000ft it might have looked just like a deliberate, if rather untidy, roll off the top.

That instinctive application of hard opposite rudder had been made into a near-reflex action many years before. Like everyone else I had been taught how to deal with a spin before going solo, but it was only later – when aeroplanes began to mean something as controllable, almost sentient things – that it became something to be mastered as a deliberately fostered manoeuvre.

RIGHT: *The Avro 631 Trainer Cadet was, for many amateur pilots of the 1930s, a nearly "grown-up" aeroplane.*

ABOVE: *The cockpit of a North American Harvard.* CENTRE RIGHT: *Cockpits of the Trainer Cadet.* RIGHT: *Instructor's cockpit of a North American AT-6, the US version of the Harvard.*

FACING PAGE: *A Harvard II of Southampton University Air Squadron.*

KF910
KF910

An Avro Cadet and a keen instructor first taught me a bit more than the bare essentials of the techniques of spinning and rolling–just as they had taught me a little bit more about other basic aerobatics. I didn't really enjoy any of it, but felt it was a good thing to learn and derived a certain satisfaction from adding some rudimentary polish to manoeuvres which would otherwise have been wild cavortings.

The little enthusiasm was certainly helped along by the nature of the Cadet. That Genet Major-engined trainer of the early 1930s was, in its staggered-wing version, the baby sister of the RAF's Avro Tutor and a direct descendant of the legendary Avro 504s–the Js, the Ks and the Ns. For the amateur pilot of those days it was a dream of an aeroplane. Not only because, with its trimmer wheel, adjustable seat, wide cockpit and shoulder harness, it seemed to be a nearly grown-up aeroplane, but because it handled, felt and even sounded like one.

The Trainer Cadet was alive as you swish-tailed it in for a landing just over the boundary hedge, or put it into a tight turn and used top or bottom rudder to hold the Genet's Townend ring in the right place on the horizon as this dark line streamed by past the nose.

It also showed me–with the help of an instructor who had the right mixture of sternness and sardonic sympathy–that spins were not merely uncomfortable manoeuvres from which one needed to learn to recover, and that slow rolls could be more satisfying than the negative-*g* circumstances might initially have led one to believe them to be.

I learned, during a spin, to look over the Cadet's centre-section, to select a point on the horizon and to stop the rotation against that point. I began to learn, in a slow roll, how much forward pressure on the control column was needed (and how much physical discomfort was involved) in keeping the nose firmly above the horizon.

Nothing can be more surprising–or reassuring–than the conversational tones of an instructor's voice when, with your nostrils full of dust and the harness biting your shoulders, he asks you, with pained sadness, whether you have noticed that the horizon is no longer quite where it should be.

Although the 40-minute instructional outing in the CFS Harvard went down in my logbook as "aerobatics", it had really been circuits as well as aerobatics. And these were probably no more than a device of the CFS to see whether a prospective ATA ferry pilot could cope with such disportings–and, perhaps, to help to give him confidence in the idea that all aeroplanes are (if you try hard enough) controllable.

To us–the first of the ATA amateurs and the guinea pigs on which its whole future might depend–the Harvard seemed at first to be an almost incredibly complicated and even potentially lethal affair. Two or three of us had flown aeroplanes with one or more of the features of the Harvard–a retractable undercarriage, maybe; a constant-speed propeller, in, say, a Vega Gull; or American instrumentation in a Stinson Reliant–but none of us had had to deal with all of them at once.

The most difficult thing about the Harvard was its innate "American-ness". Even its smell was distinctively transatlantic and most of the instruments and controls were of a size, shape, colour and position which were foreign to pilots brought up on aeroplanes with a typically British, but apparently rational, disorder in the cockpit.

Even for the more experienced among us the Harvard was nothing if not a strange device. From the moment of clambering aboard–hampered by an unaccustomed parachute and harness–and preparing to start the engine it was different and somehow awkward. The control column was too far away and the rudder pedals too near. Even the inertia starter system–in which a flywheel was spun up and a clutch engaged by operating a rocking pedal–was a device from a different world.

The Harvard was thus a very valuable trainer for British pupils–not so much because it was a mildly "hot" aeroplane by the standards of those days, but because it demanded technical attention at all times. Things had to be done properly and in the right order. A "drill" was required and its routine flying and handling had necessarily to be left largely to pre-trained reflex action. It was for the Harvard that we were taught by CFS one of the school's most valuable little tricks–a kind of mnemonic for the cockpit drill. Throughout six years of war, while flying, maybe, 50 or so different aircraft types, this mnemonic worked. With a few modifications and additions it could be used successfully even for monsters like the B-24 Liberator bomber.

Even some of the pre-trained reflexes had, for the Harvard, to be modified by an overriding "super-think". An almost perfect, instinctively controlled, power-off landing would, for instance, be criticised by the instructor. Such a landing would be good only so long as the ground was exactly where you expected it to be. The Harvard's final stall was nothing if not sharp–so it was better to motor it on and to cut the power only when it was finally and firmly on three points.

Nobody could have been more shocked than I was when, after that 40-minute session–during which, for the final half-hour, the Harvard was rarely right way up for more than a minute or two–the instructor taxied it to the apron, climbed out and told me to go off and do an hour on my own.

In fact, I did only one circuit in that particular Harvard. Something went very wrong under the engine cowling and I brought it back from the tightest possible circuit to complain of a strong smell of burning. They promptly led me to another.

It was the Harvard, for which the techniques as well as the flying had to be mastered, which provided the real beginnings of a new experience–and one which was to continue to be "new" for six more years. It had taught me that aeroplanes, however complicated and superficially different, are all aeroplanes–and can be managed. □

CHAPTER THREE

Forgotten Bomber

*

BATTLE

THE MOST surprising feature of the Fairey Battle – when at last I, as an Air Transport Auxiliary pilot, came upon it in my tour of Upavon – was not the frightening size of this dun-coloured single-engined day bomber, but the fact that there was a rather pale-faced, helmeted aircraftman sitting in its back seat. He was, my accompanying Central Flying School instructor told me, going for the ride, and providing necessary ballast at the same time. We (the AC2 and I) smiled sheepishly at each other as, loaded awkwardly and incongruously with a still-unaccustomed parachute pack, I clambered up the wing into the front seat.

I hoped, for his peace of mind, that nobody had told him that I'd never flown a Battle before, or that I'd only just completed, on the previous day, my first solo flight in any kind of military-sized aeroplane – the Harvard advanced trainer. This Battle, in which no dual instruction was possible, was taking the place, in my

brief conversion curriculum at CFS, of the unit's only Hurricane which happened on that day to be unserviceable.

Maybe this was just as well. The Battle, monstrous though it seemed to be on first close acquaintance, was in practice an even gentler beast that the Hurricane later proved to be. With its bigger wing and that extra weight to be lugged off the ground by a single Rolls-Royce Merlin, the Battle gave the innocent newcomer a little more time. This was useful when dealing with an incipient take-off swing and when considering and adjusting the position of the top of the engine cowling in relation to the approaching scenery on the boundary of the shorter of the two fields which, together, formed RAF Upavon as it was in 1939.

After the so-American Harvard, with its rows of switches and colourful instruments and controls, the Battle was a strange but comfortingly English device. Little in the cockpit seemed to be logically arranged, but the various knobs and levers were all the more easily found and remembered for that reason. The presence, too, of that most blessed of inventions, the RAF's standard six-instrument blind-flying panel, simplified the initial search for other essential gauges.

Four strong memories (apart from that of fear for the safety of the human ballast) remain of that hour-long series of take-offs, circuits and landings which had to be completed successfully before I could expect to move on to a conversion to military twins. The first of these memories is of the blast of hot air and dust from the heating system as the throttle was moved through the gate to take-off power and speed was gathered. I felt certain that something had gone wrong – or that I'd left open that which should have been shut – but the scenery on the boundary was, by then, much too near for any reconsideration of cockpit drill. In

LEFT: *Battles of No* 88 *Squadron, Advanced Air Striking Force, with an escort of Curtiss Hawk* 75*A fighters of the Armée de l'Air, over France in* 1940. RIGHT: *An artist's impression of the cockpit of the Battle prototype.*

fact, as I discovered later, Battles were like that–and very comforting it was, too, when cruising along at lower five-digit altitudes.

The second memory is of the emormous drop in engine revolutions as the two-pitch propeller control was moved to "fine" after the initial climb. The silence was almost complete as the engine speed dropped and the rpm needle swung back from 2,900 to 1,500 or so. That also turned out to be entirely normal. There was, too, the visual undercarriage indicator which sensibly screamed "WHEELS" at you in brightly lit letters as the throttle was closed.

Finally there is the memory of the undercarriage, flap and bomb-door controls which sprouted together out of the cockpit floor. You tended to lose sight of the visible world after take-off while groping around for the undercarriage lever and its safety catch. After one or two attempts this action offered no great difficulty to the normally equipped pilot. But the first group of pseudo-pupils for the embryonic Air Transport Auxiliary included a very experienced pilot with only one fully operative arm. He talked his way into being allowed to show that he could fly the Battle after managing the Harvard successfully.

It was not surprising that, on his first take-off, the Battle bobbed about quite a bit. So as to get at that lever and its catch with his good hand, he had to lower the seat as far as it would go and be out of sight of land for some tens of seconds. As with many other types of aeroplane which he ferried during six later years of ATA service, he had worked out a time-and-motion system in advance.

Like the twin-engined Blenheim bomber, the Battle was designed to meet conditions of a period earlier than that in which it finally went into brief action. In 1933, when its specification was approved, the Battle had seemed to be a very good aeroplane. Its performance more or less doubled that of the Hawker Harts

and Hinds which it was to replace. But, by the time it was necessary for the Battle to go into action in 1939-40, it was hopelessly outclassed in speed and lacked necessary defensive firepower. A very large number of Battles were built – mostly by Austin Motors under the "shadow factory" scheme under which Blenheims were also built in large (and, as it happened, largely useless) numbers by Rootes at Speke Airport, Liverpool.

Hindsight is all too easy – and mass production of anything is very difficult to stop quickly without serious social and other disturbances – but I wonder now why something could not have been done about it. It should have been practical for someone to have decided on available knowledge during 1939 that the time had come to try to divert this enormous effort to something more useful and to stop the mass production of aeroplanes which, good though they were when originally developed, had obviously become easy meat for the known fighters. After brief, heroic use as a bomber, the Battle, as such, was taken out of service, but two derivatives – the dual trainer and the target tug – continued to do good work.

For us, the ferry and test pilots, the Battle provided some easy and trouble-free test outings and a few dispersal and delivery flights which helped to break the seven-day-week routines of the early war period. What, for instance, could have been nicer than the delivery of a target-towing Battle from Worcester (where a few of them were still being assembled in 1940) to Jurby in the Isle of Man? Even in that desperate year those in command of ferry pools and maintenance units had learned that an occasional brief holiday was necessary even for non-operational flight crews. So, for this trip, I and another ferry pilot had been given permission to stay in the island for 48 hours of leave before collecting an Anson to take to Hawarden, Chester, for onward delivery.

It was a lovely cloudless day in June. You could see for 50 miles or more, so I decided that there was no need to use the very shortest sea crossing from St Bees Head, Cumberland, but could compromise by pottering up to Squire's Gate, Blackpool, and setting course from there. Most of the ferry pilots were accustomed to the Irish Sea crossing from Shorts' factory in Belfast to maintenance units in various parts of England and from factories in England to the maintenance unit at Aldergrove in Northern Ireland. But those flights had been made with twin-engined aeroplanes like Hereford bombers from Belfast and Ansons from the Avro plant at Woodford, Cheshire.

As the Battle cruised over the Lancashire conurbations (if I may use that modern and overworked word) towards the Ribble Estuary there seemed to be an awful lot of sea over to the north-west. Even under that cloudless sky it looked limitless and very wet. Nervously, I climbed the Battle up to a dizzy 10,000ft, took several deep breaths of the thinner air, turned the compass verge ring to 310 degrees magnetic, settled on to course and re-set the gyro to zero. Why is it that powerplants always run roughly and engine instruments always flicker when over the ocean? I was comforted by the sight of the Cumberland coast which, for a while, stayed more or less the same distance away on the right. Presently Snaefell, topped by its usual little cloud, could be seen over the engine cowling.

Not much more than two hours later – rid of the Battle and of such things as a parachute pack, flying overalls and the thought of the next ferrying job – I was looking at the boats in Peel harbour and chatting to my companion ferry pilot about peacetime pleasures. We felt innocent enough, but we must have seemed – in our old blazers and flannels – to be fairly suspicious characters as we stared at those boats on an island where enemy civilians were interned. The two policemen who closed in on our right and left were not too quickly convinced of our innocence.

Some later delivery and dispersal flights with Battles were not necessarily so relaxing. By the mid-war years the remaining Battles – except for the target tugs and the few dual trainers which had not been shipped to Canada – had become a nuisance to the units which stored, maintained and dispersed them. The maintenance work, often done at the cost of more important duties, became fairly minimal. The MU test pilots – who, when they could spare the time, shuttled them about the country to and from dispersal fields – found that they had to put up with such things as dubiously effective hydraulic systems and inadequate braking power. The Battles were all made to work after a fashion – that was an essential – but the test pilots learned not to be too fussy when they knew that there were much more important aeroplanes to be looked after.

For some long-forgotten reason I ferried several Battles to Barton aerodrome, near Manchester – a small grass field (once Manchester's airport) which I had known quite well in earlier years. One of these Battles had extremely feeble brakes and could not be encouraged to deliver more than about half the maximum flap extension. I made three approaches before I managed to get the speed down low enough for the Battle to be landed firmly just inside the boundary. Even then I had to organise a mild ground-loop, using the remains of the differential braking and a burst of power, to prevent an uncomfortable plunge over an embankment.

I still find it difficult to imagine what happened to all those thousands of aeroplanes which were built during the period of hurried rearmament and which became redundant almost as soon as the shooting war started. Of the Battles, some were left in France in 1940 and the target tugs and trainers continued to do useful work. But many more than 2,000 were built and the remainder must have been parked *somewhere*. How often do you see a dead bird except when it had been hit by motor traffic or killed by a domestic cat? Maybe aeroplanes, too, are quietly translated to some mysterious, mechanised Valhalla.

The Battle, that gentle old tin Swallow,* deserved better than to moulder away in some forgotten dispersal field. □

**The pre-*1939 *Swallow was a big-span, low-wing two-seater light aeroplane developed by British Aircraft from the German Klemm. The Battle behaved very much like it when you were ready to forget that it had ten times more power and was a lot bigger and heavier.*

CHAPTER FOUR

Life-saving Drill

*

BLENHEIM

JUST WHEN the Bristol Blenheim reached its engine-failure safety speed after take-off there was a sudden change in the sing-song clamour from its two Bristol Mercury radials and I found myself instinctively holding heavy pressure on the left rudder pedal to keep the outfit straight. The starboard engine had failed.

Beyond the boundary of the grass airfield lay all the paraphernalia of a large industrial city–factories, gasholders and chimneys. Well, plenty of other people had taken a Blenheim round successfully and back to the aerodrome on one engine, so I supposed it could be done.

During that terrible winter of 1939–40, with six months still to pass before the end of the "phoney" war and the surge of German armour into the Low Countries and France, the ferry pilots, civilian and Service, were working through the short daylight hours–and sometimes into the night. They were delivering thousands of aircraft–few of which were ever to be flown in hot blood–from factories to RAF maintenance units for storage and for arming, and, sometimes, from MUs to operational squadrons.

This Bristol Blenheim Mark IV was due to be taken from Speke aerodrome, Liverpool–where the Rootes shadow factory was mass-producing these light bombers–to the maintenance unit at Hullavington in the Cotswolds. Now it looked as if there might be more than a fifty-fifty chance that it would end up, the worse for wear, in a city street.

Flying at chimney height on the circuit, and trying to remember all the things a virtuous Blenheim pilot should and should not do to bring himself back alive on one engine, I was much too busy to consider the prospects of failure–but glad enough when the grass was once again brushing under the wheels for a landing rather far up the field.

This sort of thing was, after all, what we in the Air Transport Auxiliary had learned to cope with during our short course at the RAF's Central Flying School in September 1939. After a couple of hours or so on the Harvard advanced trainer and an hour in a Battle light bomber, on which no dual instruction was possible, some of us were led to a Blenheim to see if we could cope with this twin-engined monster.

The Blenheim was a small aeroplane by present-day military standards – with, maybe, no more than twice the weight and three times the power of a typical executive twin such as the Beech Baron – but it was a hot affair by the standards of those days and quite a few pilots had been killed through lack of understanding of engine-failure handling techniques. Once mastered, it was a beautiful aeroplane to fly. In the Mark I version, the pilot sat on a kind of dais in a glasshouse; he could see ahead, below and nearly all around him except where the two engines sprouted out of the wing.

When, in my turn, I had been shown around the Blenheim's alarmingly complicated control cabin, the CFS instructor airily waved me into the left-hand seat and settled himself down beside me in a seat where the dual-instruction arrangements were, to say the least of it, somewhat primitive. Did the instructor know that I'd never flown anything like this before? He did. "There is nothing to it," he said, and began to go through the engine-starting drill. "I'll do a circuit first and then you can have a try."

Ten minutes later he had folded his arms in his lap while I sweated my way through the take-off drill with the Blenheim sitting across wind at the leeward end of Upavon's small grass airfield. I had already learned the basic mnemonics by which the CFS instructors taught their pupils to remember the essential actions. For take-off the letters were HTMPFG – which I remembered by muttering "hot-tempered Member of Parliament flaps his gills" – and for the final approach drill they were HUMPFG, or just HUMP. Hydraulics, trimmers (and throttle tighteners), mixture, propeller pitch, flaps (and fuel cocks) and cooling gills (or radiator shutters) – these were the essential things on take-off. The "U" in the approach mnemonic stood, importantly enough, for undercarriage.

LEFT: *Cockpit of a Blenheim Mk I.*
RIGHT: *The Bristol* 142 *"Britain First" – the prototype for the Blenheim.*
BELOW: *Blenheim Mk IV of No* 53 *Squadron.*

The Blenheim had a pretty odd control layout. On his right the pilot had three little handled plungers–rather like the things in old–fashioned lavatories–and these were pulled up or pushed down according to requirements. That at the back was the master hydraulic control, providing pressure to the main systems or to the retractable gun turret as required; the other two operated the undercarriage and the flaps respectively. The mixture control, beside the throttles on the pilot's left, had three positions–automatic weak, rich and, dramatically enough, take-off override. The two-pitch propellers were controlled by a couple of fore-and-aft moving plungers in a panel on the left and *behind* the pilot; with them (luckily protected by a hinged cover) were similar plungers to operate the engine cut-outs. The cooling gills were operated mechanically by handwheel.

Well, there I was, for better or for worse. Hydraulics selected, rudder and elevator trim adjusted, throttles tightened, mixture in override, pitch checked in fine, flaps up, tanks selected and gills wound closed and then opened a trifle. I turned the Blenheim round into wind and started, gingerly enough, to open the throttles.

I need not have worried. The Blenheim went quite straight without any throttle differential and seconds later I had lifted it off the ground, re-trimmed and pulled up the undercarriage plunger–losing a length of thumb-skin in the process. "Hold her down", said the instructor, "and keep holding her down until you've got engine-failure safety speed." At an indicated 115mph I pulled the Blenheim up in a climbing turn and reached behind me for the propeller-pitch knobs.

After cruising around for a few brief minutes of familiarisation, the Blenheim was brought into the circuit and I started to go through the approach drill.

Hydraulics checked; undercarriage plunger pushed down and green lights flickering on; mixture in override; pitch-knobs hammered into fine, the two Mercury engines screaming in turn as the revolutions increased; and the flap plunger pushed and pulled quickly back into neutral as the indicator showed the correct approach setting.

As the boundary fence came up, with the Blenheim floating along on a whiff of throttle, the flap plunger was slammed down and left down. As the big split flaps extended, the aircraft felt as if it had hit an arrestor wire. But we had plenty of speed and the throttles could be cut half-way through the flare-out. The Blenheim settled and stayed settled and an occasional jab of the thumb on the brake lever served to keep it straight.

As I held the Blenheim down on my third take-off I felt one of the throttles being pulled back and saw a chamois-gloved hand reaching for the switches. The instructor had cut the starboard engine. "Keep her straight", he said, "and hold her down; don't start to climb yet." The technique of engine-failure safety speed had been invented.

And so it was, at Speke a few months later, that there was nothing so very unusual or even frightening about the situation when the starboard engine banged and popped itself into final quietness. I had been taught the trick of gaining all possible speed before reaching the boundary and the only difference was that, instead of the farms and trees of Wiltshire flickering by a few feet below, there were houses and factories below and gasholders and chimneys seemingly above my flight level.

I edged the Blenheim round, with the good engine on the inside of the turn, praying that this engine would stay good until I'd got well along the downwind leg of the circuit and safely over the Mersey Estuary. It kept going.

CHAPTER FIVE

No Riotous Reception

*

ATA

A GROUP of prospectively undisciplined and uniquely uniformed strangers could hardly have expected a riotous reception at any RAF unit in 1939. At the RAF's No 2 Ferry Pilot's Pool, Filton, Bristol, the Air Transport Auxiliary pilots were, at first, little more than tolerated as intruding civilians and were made to feel very much on trial–with failure expected.

Our private lives–such as remained to us–were not so displeasing. The Bristol citizens, on whom the ATA had been forcibly billeted, bore with us well and one or two of the more independent pilots leased flats or houses and so contrived a return to normality when and if a day's work was over. Although we enjoyed this limited freedom from some of the minor indignities of Service life, that first period of the ATA's existence, while its pilots were learning something of the ferrying business, was very hard–and not only because of the very difficult 1939–40 winter weather.

We were continuing to learn to fly strange and complicated aeroplanes, yet there was no time for learning. Once airborne we had to get aircraft from A to B as briskly as possible or risk being stranded by the impatient pilots and passengers in the taxi Ansons. Our uniforms were strange and nobody knew who or what we were; RAF adjutants had to be cajoled, and sometimes actively bullied, into giving us transport to the nearest station and railway travel warrants when strandings inevitably did occur. Sometimes we were away for days at a time and had to buy shirts and underclothes as we went along.

But all the time experience was being built up by those who had got through the RAF Central Flying School courses and every day the deliveries were less liable to end in a disaster.

So when, in February 1940, the first independent ATA pool was formed at White Waltham, near Maidenhead, there was a nucleus of reasonably experienced ferry pilots to keep things going in a situation which was, through no fault of the sponsors and organisers, as nearly chaotic as made no matter. The ATA had a base and the beginnings of a ferry programme–but, because it was a civilian organisation, it had, initially, no direct power to obtain, requisition, or order such important things as taxi aircraft (essential for getting the pilots to the aircraft factories and from the RAF units) and parachutes (essential for sitting upon in bucket seats, even though no one expected seriously to have to use them for their survival purposes).

The parachute and other similar equipment problems were solved by the simple expedient of arranging for the secondment of a trained RAF equipment officer – who was, fortunately, a volunteer reservist with no rigid Service ideas and a great wish to get the ATA going. He knew the form-filling and other routines and was wearing the right uniform, so he commandeered transport and just got the necessary items from RAF stores.

As for taxi aircraft, we had initially the most extraordinary assortment of cast-off RAF and "impressed" civil aircraft. At the top of the list were two of the oldest of Avro Ansons (Mark Minus Ones, we called them) with no flaps. These were supported by a few American Stinson Reliants of various and inevitably ancient vintages, some Airspeed Couriers with their undercarriages permanently locked down, and odds and ends such as a DH Puss Moth or two.

In fact, we got the jobs done, as often as not, by using as taxis the aircraft we were supposed to be delivering. An Airspeed Oxford trainer or a Percival Proctor might, for instance, travel by the most involved and circuitous route (including one or more night stops) during delivery from its manufacturers to the maintenance, training, or other unit named on our delivery chit. All this had to be planned with meticulous, yet commonsense, precision so that as few pilots as possible were left to return to base by a time-wasting (and then excruciatingly uncomfortable) train journey.

By the spring of 1940 it had been decided that the ATA should eventually be responsible for all ferrying work except that to and from foreign parts and in May I was given the job of opening up the first branch establishment at Hawarden, Chester. At about the same time another sub-pool was opened at Whitchurch, Bristol, and later the next civilian pool was opened at Prestwick, Ayrshire. Before the end of 1942 there were 16 individual civil ferry pools. In the late summer of 1940, however, I had succeeded in a second attempt to get past the RAF's medical and selection boards and had, with regrettably underhand lobbying, got myself posted to a Service maintenance unit as a test pilot. By a lucky chance there was an establishment vacancy in the test flight at Hawarden, and to this I found myself posted. So all I had to do was to walk round the perimeter track from No 3 FPP, ATA, to No 48 MU, RAF Hawarden.

Things were not always quite so simple for individuals in that first full year of war.

The year with the ATA, first on attachment with the RAF and later with the ATA's own ferry pools, provided essential experience and training; many things happened, pleasant and unpleasant. After the successful delivery of Blenheims (which the CFS, after all, had said some of us were capable of flying) the CO of the RAF ferry pool to which I was attached arranged for me to have some dual instruction on an Avro Anson and let me fly it on the occasional taxi operation – but the first unfamiliar "hot" aeroplane which I was allowed to deliver was a Hawker Hurricane. Later, as a test pilot with the RAF, I was to spend a lot of time in the air with this strong and forgiving fighter.

CHAPTER SIX

The Gentle Hurricane

*

HURRICANE

THE LITTLE MOUNTAINS of Snowdonia were iridescent in the dappled sunshine of an evening in the early summer of 1942. It was good to be alive–and especially good to be cruising quietly over North Wales in such a familiar aeroplane and on such a relatively pleasant job of work. For once, there was time to look about and to think.

Periodically during those early war years maintenance unit test pilots would take Mark II Hurricane fighters on simple little fuel-consumption tests and this flight was one of them. There had, so the grapevine of those security ridden days explained it, been some trouble on a mass delivery flight to Malta or North Africa from an aircraft carrier in the Mediterranean, and it was our job to see that the failure would not be repeated.

The method used was primitive enough, but it worked. The Hurricane was taken off on fuel from the fuselage tank, and any minor outstanding test checks were completed. Boost was reduced to plus two and revolutions to 2,000; the Hurricane was then settled into a level cruise at exactly 2,000ft, the fuel supply was switched over to the wing tanks and a (carefully preserved and individually issued) stop-watch would be clicked as the fuel cock was turned.

At the end of exactly 60 minutes the supply was switched back to the fuselage tank and the Hurricane was brought in to a landing. The amount of fuel needed to top up the wing tanks was then measured. It was as simple as that.

The system was not entirely fool-proof, but as nearly so as made no matter. The enjoyment of the panoramas of Snowdonia (or of the Pennines, or of the borders of the Lake District) could not be permitted to distract attention for too long from the job in hand. The test would be washed out if the chosen route, or the weather, required any change to be made to the cruising level or power. It was also necessary to judge the duration of the tour so that the aircraft would be back over base when the hour's flight had been completed and the engine was again being fed from what fuel was left in the small-capacity fuselage tank. In earlier days we had cruised steadily around our aerodrome, but we very soon became bored with that unimaginative method of completing the exercise.

The Hurricane had, by 1942, become so familiar to the pilots of our unit that it would be taken on flights to check the weather, when this was marginal, in preference to any Tiger Moths or Magisters which might be lying around. After

250 hours or more of experience, divided into a series of 15-minute test flights, any aeroplane becomes almost a living extension of one's hands and feet.

A conscious effort was needed to remember the time, much more than two years before, in November 1939, when, stomach revolting, I had climbed aboard one of these fighters for the first time and tried my best to look confident enough to convince the manufacturer's ground crew that the thing would be moderately safe with me on its delivery flight.

The gentle Hurricane had seemed then to be a stark, brutal and altogether terrifying device as I looked out over the long engine cowling whilst absent-mindedly trying, with my right hand, to devise a way in which the safety catch in the undercarriage-cum-flap-lever gate could be moved quickly out of the way when the time came, after take-off, to select "up". There was, too, in those early variants, an hydraulic-power lever which had to be held down during undercarriage and flap operation and I had only two hands.

Everything went reasonably well on that first trip, though I suffered a heart-stopping fright when the untended throttle lever vibrated into a near-closed position while I was holding down that hydraulic lever after take-off. I had, too, fought quite unnecessarily with another lever which turned out afterwards to be one for two-position pitch control and not, as I thought, for constant-speed propeller adjustment.

Slowly over the months the Hurricane had become a friendly old war-horse, but, inevitably at any maintenance unit, there were occasional unpleasantnesses. Nothing as complicated as an aeroplane, even one of that era, could be expected to function perfectly after storage and modification work in a relatively primitive environment.

Gun panels were sometimes shed in flight, leaving on the upper surface of the wing a gaping hole which, even in our aerodynamically limited experience, was very obviously going to affect the stalling characteristics to no small extent. When that happened, the Hurricane would be wheeled on with plenty of excess speed and the pious hope that it could be stopped before the end of our short temporary runway.

Very occasionally, too, even a Merlin engine might show signs of the beginnings of some internal malaise. Then, seeing the rising oil and coolant temperatures and the dropping oil pressure, we would park the thing back on the ground as quickly as possible, leaving the engineers to search the filters for signs of gold dust.

I can remember only one really unpleasant test-flying experience with a Hurricane and even that had been caused by over-zealous wartime effort. We had been having a certain amount of bother at my unit with undercarriages which would not lock up. I had taken off for a test flight after all-night work on a particularly troublesome Hurricane. Everything had gone well until I had tried to get the undercarriage down. Successive selections of gear and flaps and vigorous efforts with the emergency hand pump had done no good. There was a shortage of green lights, the warning horn blared when the throttle was closed and it was obvious, from the behaviour of the aeroplane, that one undercarriage leg was staying up whilst the other was dangling in the slipstream.

Cruising around, I thought hard about the situation. This, perhaps, was the

occasion which would justify the use of the tatty parachute which had for so long served as the parody of a seat-cushion. I pulled the pin of my Sutton harness, slid back the canopy and considered, not too seriously, the various approved methods of abandoning this particular ship. Then, angered by all the implications of a ridiculous situation, I made one final attack with the hand pump as this became "solid" after the dangling leg had, once again, been the only one to lock itself down.

Suddenly the resistance eased, the hand pump moved freely and both green undercarriage lights came up bright and infinitely cheering. Unencumbered by the safety harness and the canopy, I had been able, almost standing up in the cockpit, to apply the strength needed to give enough hydraulic power to the jack and to break the port undercarriage leg free.

In fact, as it turned out later, the locking stirrup of this leg had got itself wedged behind the undercarriage-well cover-plate; that on the other leg had broken through its plate and would not go back. Hence the one-up-one-down situation. In the hurry at the end of the all-night session nobody had remembered to do a final retraction test after the two cover-plates had been put back. Safely on the ground again, I was in no mood to complain too vigorously about the little lapse.

□□□□□

Looking at the stop-watch, I saw that there was not much more time to be killed on this consumption test. Turning the Hurricane over that decapitated hill behind Penmaenmawr, I started to follow the coast back to base. The sun was only just dropping below the horizon, and there was still plenty of light at 2,000ft but the landmarks below were beginning to fade and night flying was not on our agenda.

Ten minutes later, the test completed and the fuel supply switched over, I was glad enough to find that the Spitfire operational training unit at my station had laid down a flare-path. The Hurricane touched gently in a satisfying three-pointer as I watched, out of the corners of my eyes, the second pair of flaming goosenecks go by. It had been a long day and I hoped that the mess would be able to raise some sandwiches. □

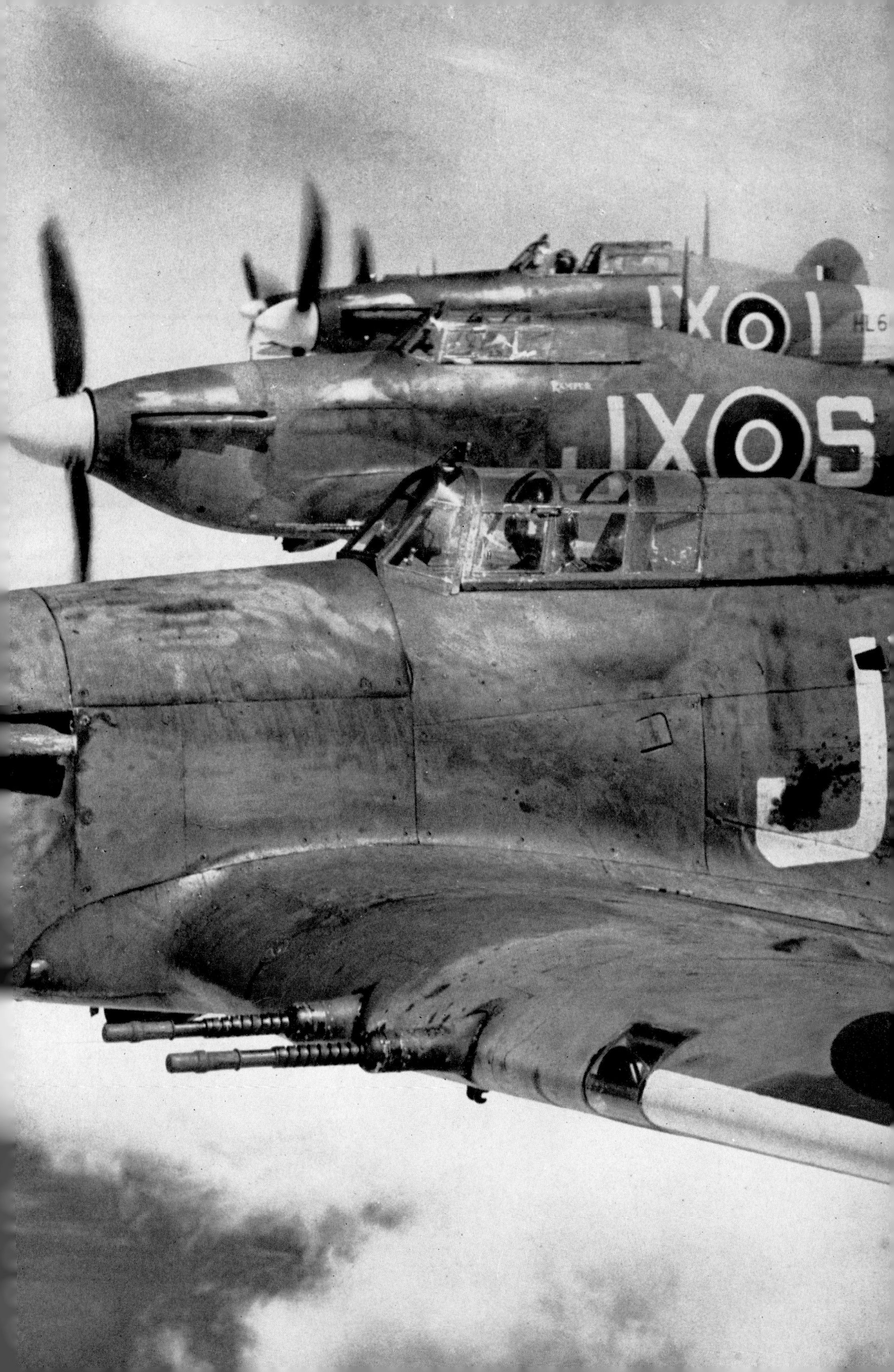

CHAPTER SEVEN

Last of the Biplanes

*

SWORDFISH · GLADIATOR · GAUNTLET

THIS RIDICULOUS SITUATION had been met at least once before and would be met again many times more in one form or another. Here I was, airborne in an entirely unfamiliar aircraft, and finding that I had either failed to do my homework properly or had forgotten what I had learned. For five minutes now I had been racking my memory and searching the cockpit in an unsuccessful attempt to remember the whereabouts of an absolutely straightforward instrument–the fuel gauge or gauges. No such thing was to be found anywhere in the reasonably spacious and uncluttered cockpit. Such failures of memory or briefing, of even greater importance than this, were to be expected (though hardly to be approved) when one was ferrying two or three different types of aircraft each day.

The month was November and the year was 1939–logbook-recorded trivia which, nevertheless, have some importance in the narrative. For six weeks or so I had been attached, with other members of the then-embryonic Air Transport Auxiliary, to the RAF's No 2 Ferry Pilot's Pool at Filton, Bristol. For some forgotten reason the usually strict and conformist commanding officer had relaxed on this particular day to the extent of letting me choose, from the daily ferrying movements list, the aircraft which I would prefer to fly. It was, in fact, my birthday; but there is no reason to suppose that he knew this, or that the knowledge would have led to any relaxation of attitude. The ATA was there to do a job of work–and that was all.

There were, on the list, the usual crop of long-nosed Blenheim IVs from the Rootes shadow factory at Speke, Liverpool; a Vickers Wellington or two from Brooklands; a Handley Page Hampden from Radlett; and some oddments. I chose one of the oddments–a Fairey Swordfish which had to be moved from the nearby maintenance unit at RAF Kemble in the Cotswolds to the MU at RAF Sealand, near Chester. To get back, there was one of the inevitable return ferry trips bringing a Blenheim from Liverpool to the MU at RAF Hullavington, where I would be picked up again by an Anson and flown back to base. This gave me a relatively easy and straightforward day's work.

The Swordfish was chosen because it seemed to me to be a nice, simple biplane–little more than a very grown-up version of the DH Cirrus Moth in which I had learned to fly ten years before. So I could relax and, just for once, enjoy flying for its own sake after six weeks of prostrating daily struggles with more complicated devices and their involved cockpit drills.

Torpedo-spotter-reconnaissance Swordfish from Gosport, practising dive-bombing peel-off in 1939.

So here I was, cruising over the south-west Midlands in this kindly old biplane, not knowing how much fuel there was on board and too proud to go back to Kemble to be told where the gauge could be found. In any case, I was reasonably relaxed in the conviction that the Swordfish could, if the engine stopped, be sideslipped and swish-tailed without too much difficulty into any good-sized field.

In the event the Bristol Pegasus radial and its fixed-pitch propeller tick-tocked along without hesitation all the way to Sealand. I found out afterwards that the gauges for the main (wing) and gravity (fuselage) tanks were behind the instrument panel and could just be read (if the acetate windows above were clean enough) through a hole above this panel. Another oddity about the Swordfish was the hand-sized, knurled wheel at the rear of the centre-section trailing-edge. This lowered the ailerons symmetrically to provide a partial flap effect.

In retrospect the choice of a draughty old biplane for the first delivery of the day is still something of a mystery. Relatively elderly though some of us were in the ATA of that era, we were all keen to get on to new types. By then I had done my stint of Blenheim IV deliveries (the RAF's Central Flying School had passed me out on a short-nosed Mk I, which was the only twin available at Upavon for dual instruction during the brief conversion course), flown my first Hurricane and delivered a once-and-for-all Vickers Wellesley to its final resting-place. I had also been for look-see rides as a passenger in Wellingtons, Whitleys and Hampdens and was ready to have a go at almost anything. So why did I pick a Swordfish?

The real reason was not so much that I wanted a slow, safe ride. I knew that monoplanes with retractable undercarriages were likely to be much safer in a forced landing than any fixed-undercarriage biplane. But I wanted, just for once, to be able to avoid the need for the life-or-death concentration which the more modern aeroplanes demanded in the way of cockpit checks. I was tired of mumbling desperately through the drill before each take-off and landing, while praying that nothing vital–gills not closed, perhaps, or propellers not in fine pitch–had been forgotten. Trim and mixture settings were about all you had to worry about in any of the old biplanes. Nevertheless the illusion of their greater safety remained throughout the war and the offers of a delivery or test flight of a Swordfish, a rare Gloster Gladiator, or still rarer Gauntlet, were accepted with relaxed pleasure.

My one and only experience with a Gauntlet, the two-bay biplane predecessor of the Gladiator, hangs in my memory because its delightful handling characteristics were overlaid by physical discomfort and because the starting procedures for the earlier version of the Bristol Mercury radial were very complex. This ancient and battered Gauntlet had to be delivered in the spring of 1940 from the Royal Aircraft Establishment, Farnborough, where it was no longer wanted, to the MU at RAF Aston Down, where it no doubt rotted quietly away.

The discomfort stemmed from the facts that the Aldis gunsight (a telescope-like affair which protruded through the windscreen towards the eye of the pilot) had been removed and that the hole in the windscreen had not afterwards been covered. The result was that a blast of cold and oil-impregnated air was directed at my face throughout the 40-minute delivery flight. The jet of air periodically got under my goggles and lifted them off my eyes; when they were pushed up on to the forehead the blast was not only unacceptably uncomfortable, but the goggles felt as if they were floating about on their elastic. In the end I pulled them down to my neck and avoided the blast as well as I could by moving my head to one side or the other.

Gladiator delivery flights had provided a somewhat different experience. These were brand-new aircraft from the Gloster factory at Brockworth and had only just completed their production tests. So everything worked to perfection; there was little risk of trouble of any kind and a near-perfect flying experience was provided for the ferry pilot who was lucky enough to be put down to deliver one.

The first Gladiator I flew had to be taken from Filton – where it had been parked for the night after being forced back by weather on an earlier delivery attempt – to the MU at RAF Shawbury on a brilliant winter day late in 1939. When the hood

The reasonably spacious and uncluttered cockpit of the Swordfish.

was wound shut the cockpit was more than a little cosy, if somewhat smelly. One ignored the fact that, by newer standards, the pilot's view was somewhat impeded by the centre-section and assorted struttage.

The track took the Gladiator over the Forest of Dean, to the west of the Malvern Hills, over Wenlock Edge and past the Wrekin to the plains of northern Shropshire. The flight was made at 800ft or so all the way, with occasional minor deviations so that the handling in tight turns and other mildly off-beat manoeuvres could be enjoyed. I had been warned that, if allowed to do so, the engine stopped all too readily after the landing. The Gladiator was floated into Shawbury (then a grass field) with the throttle closed and a comfortable amount of glide-steepening drag from the little flap-type air brakes. As it was being held off for the landing I was suddenly conscious of an all-too-slowly flailing propeller and caught it just in time with a burst of power at the cost of a somewhat untidy landing.

One of the oddest things about those early months of the war was the way in which the ex-amateur ferry pilots, some of whom had never flown anything larger or more powerful than a Moth Major or Hornet Moth, were able, if they had been passed out by the CFS, to take the flying of these bigger military types

RIGHT: *"Stringbag" the sailor – a Swordfish complete with torpedo.* BELOW: *Still in production in 1939, the Gladiator was the ultimate in biplane fighters.* BELOW RIGHT: *Predecessors of the Gladiators – Gauntlets of No 56 Fighter Squadron.*

so very much for granted. Three months before, nobody would have let me loose on a DH89 Rapide (or Dominie as the RAF called it), even if I'd had the confidence to attempt to fly one. Yet this little twin-engined airliner was considered to be simplicity itself and a delivery from Hatfield to, say, the school at Yatesbury, near Chippenham, was looked upon as a very gentle chore.

When struggling for most of the time with much more fearsome devices, it was necessary to remember that the kind of aeroplanes on which we had been brought up had their own difficulties and could turn and bite the unwary. In some ways the monsters were easier to fly because of their high wing loadings and ample power.

The lesson had to be re-learned by experience from time to time. Some three years later, when in the RAF on test-flying work, for instance, a Tiger Moth came up for check among the widely assorted twins and multis prepared by the unit to which I had been posted. Months had gone by since I had handled anything lighter than, say, a Hurricane or Miles Master, and I had forgotten that a Tiger really had to be flown off and on to the ground–with repayment in the form of the immense and long-lost satisfaction provided by a rare thistledown three-pointer.

After a pleasant half-hour of aerobatic practice the Tiger was brought in to land on a runway which was slightly off-wind. Carelessly, I let the wheels touch prematurely and the Tiger proceeded along most of the length of the runway in a series of bounces and short power-corrected flights before I decided that it would be better for all concerned if I called it a day and went around again for a new attempt. This time the Tiger settled on three points with the drift accurately kicked off, but my face was still red–and not only from the cold air of that December day in 1942.

□□□□□

Within a week of that first restful Swordfish ride I found my name down to deliver a Handley Page Hampden medium bomber from Radlett to an MU in the Midlands. It presented no particular problems, even to a beginner on military twins–but the story of the Hampden fits in better with later test flying experiences. Not many days later, after struggling successfully with a Hawker Henley among by-then better-understood types, a little group among the small ATA contingent at Filton were sent to collect and deliver Westland Lysanders from Yeovil. □

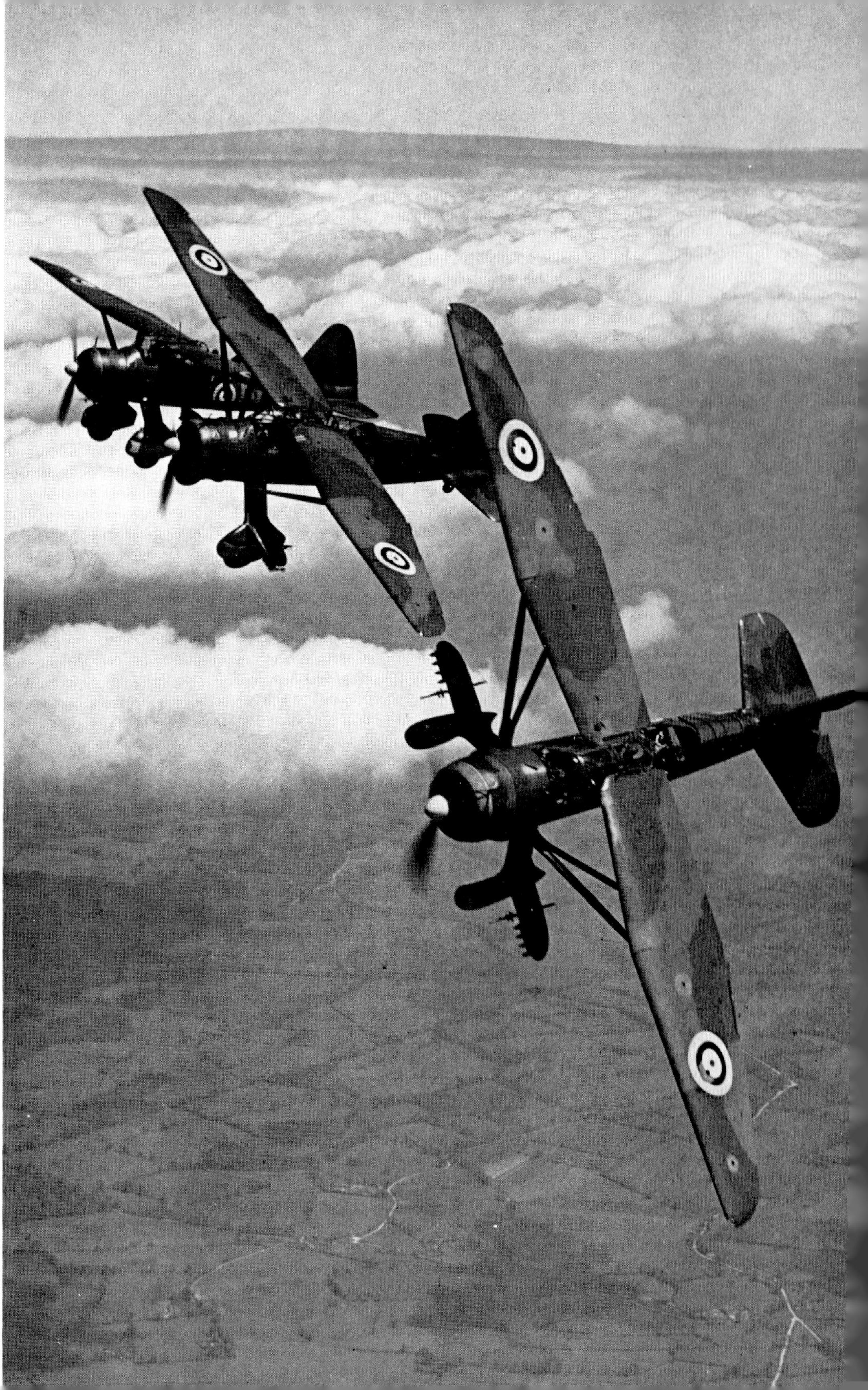

CHAPTER EIGHT

Cloak and Dagger Aeroplane

*

LYSANDER

THE TWO OTHER PILOTS on this little ferry outing may or may not have been accepting the situation with appropriate British phlegm (whatever that may be) – but I was having kittens as I tugged at the ends of the Sutton harness. This, P9077, was my third Lysander so I was familiar enough with the type and there was no normal reason for my fit of nerves – which was caused by the sudden appearance of a little wall of mist creeping over Yeovil's grass airfield.

Our taxi Anson had already taken off to complete its complicated rounds before picking us up at Cosford, where the three Lysanders had to be delivered to No 9 Maintenance Unit. The other pilots were already turning into wind for take-off – and here was I, tail-end Charlie as usual, left with what was likely to become a logically insoluble weather problem.

Well, I would take my time over the engine check, taxi out to the leeward boundary and then have another good look at the situation before making up my mind. A delayed decision would at least leave me free to concentrate on the engine run-up and the preparation-for-take-off drill.

The Lysander was a strange, but (even for amateur ferry pilots) a relatively unworrying device. It had only one engine and a simple two-pitch propeller; it was comparatively easy to conduct in normal conditions; it gave its pilot an extraordinarily good view of the world at large; and it could (so the experts told us) be landed on a cricket pitch.

Some of us were to find, in due course, that the Lysander's simplicity could be complicated. There was, for instance, the need to remember the word TRIM. Success in autogyropic arrivals was the prerogative of the specialist; and it could demonstrate one or two uncomfortable, if fairly harmless, characteristics during an ill-executed low-speed final approach.

On that day in January 1940, I had little more than two hours of Lysander flying in my logbook. During the previous delivery flights I had treated it as a conventional aeroplane and, in return, it had behaved like one. After the struggles with the more complicated aeroplanes – such as Blenheims and Hampdens – which appeared on our daily ferry programmes, it was a relaxing affair for the inexperienced ferry pilot of this era.

But there was no feeling of relaxation for me this time as I turned the Lysander into wind and studied that amorphous white mass of vapour as it crept over the aerodrome.

For a few seconds I considered the pros and cons of the situation. The Midlands and North were clear of fog and the forecast had said that they would remain clear at least until the late afternoon. There had been no ground mist to be seen from the ferry Anson until well south of Bristol on the way to Yeovil from Filton.

Once airborne, therefore, the Lysander had to keep going for no more than a quarter of an hour before the ground would come into view again–and there were still quite a few hundred yards of aerodrome to be seen on my side of the mist. But it was now or never.

I thought of the long, tedious journey by train and taxi back to our Filton base–and opened the throttle.

The Lysander was airborne long before reaching the mist and touched only a few wisps of it as we climbed away. The view, as I turned on to the heading for Cosford, was quite phenomenal. All around was a brilliant white sea with the hill-tops poking through it like little spinneyed islets. The Bristol Perseus engine was running like a sewing machine and soon I could see the Mendips to the north and the Bristol Channel to the north-west. I was almost sorry when the mist began to clear down below.

Later, much later, I learned a few of the tricks which were possible with this unique device when it was not being treated as a conventional aeroplane–and also about the characteristics which were happily accepted in payment for its ability to imitate the Autogiro.

For instance, there was that little matter of trim adjustment. The Lysander could not be held down at take-off power for an overshoot if the trimming wheel had been wound back for a normal engine-off tail-down landing. On a go-around the trick was to inch the throttle open in penny numbers while winding forward as furiously as possible on the big trimming wheel. If, by horrible mischance, you were so ill-prepared as to fail to re-trim before take-off you were in something of a difficulty unless and until you realised that the Lysander would fly quite happily on very little power while your throttle hand found time to go to work on that trimmer.

The Central Fying School-converted ferry and test pilots can rarely have got themselves into that kind of trouble with the Lysander–or with any other aeroplane. Forgetfulness on take-off was inhibited by the familiar HTMPFG mnemonic. The words "hydraulics, trim, mixture, propeller, flaps/fuel and gills" were always shrieking through our dumb minds even in the most harassing of take-off circumstances.

But I did once find myself in trim trouble with the Lysander–trouble against which normal training was not much of a defence. Someone forgot to remove the tail ballast when one of the riggers at our maintenance unit decided, at the last moment, to come along as a passenger on a test flight–and I omitted to check that it had been removed.

Even with full forward trim and with power reduced to a minimum, a great deal of strength had to be used on the stick to keep the nose down. Thething was

sitting on its tail at a hundred feet before I could get the trimmer rolled forward and the power reduced; by then I was committed to a circuit.

My logbook says that the time for this flight was 15 minutes. It wasn't–but this (following the convention of our unit's flight records) was the shortest acceptable period of time; without such a limitation a fearful succession of odd numbers of minutes might, we feared, pester us for our remaining years (months, weeks ?) of logbook additions and computations.

Such minor events–which included the shedding of inspection panels and similar incidents–remained unrecorded in the remarks column of our logbooks. After all, these solemnly numbered Form 414s were examined at the end of each month by the unit's CO and there was no point in inviting troublesome explanations. The engineer officers took these incidents seriously enough and applied the necessary corrective measures.

As for the cricket-pitch landings, even the experts were very occasionally defeated if the Lysander's Mercury or Perseus engine chose to snort loudly, instead of delivering, when the throttle had to be opened rather quickly during the final stages of a spot-landing descent.

Slots and slot-operated flaps notwithstanding, the air just couldn't, without a slipstream, hold the Lysander up when it was pretending to be a parachute. More than one specialist was left sitting idiotically in an otherwise undamaged fuselage-cum-undercarriage after the engine had fallen off during an arrival of unplanned suddenness.

Meanwhile, the happy non-specialists had found the Lysander to be, when treated normally, a most restful and pleasant affair. The view, as I've said before, was marvellous; the pilot sat high up, looking over the top of the engine cowlings;

and the deep side windows of the cockpit could, in hot weather, be slid right down. One was then poised on a cool–if draughty and noisy–dais from which the familiar world looked horrifyingly real to any vertigo-liable person.

Only when treated carelessly did the Lysander behave uncomfortably. There was, for instance, its rocking-horse game. The Lysander had full-span slots; the outer sections opened and closed freely, but the inner sections were mechanically connected to the flaps.

So a quaint situation could be allowed to develop. A reduction in speed below about 90mph opened the inner slots and lowered the flaps. These altered the trim so that the nose dropped and speed increased–thus closing the slots and raising the flaps. The nose would then try to go up, the speed would be reduced, the slots would open and the flaps would come down . . .

We–the Lysander's Service test and ferry pilots–were no more than playboys messing about with characteristics which were, later in the war, to be used for some fairly strange cloak-and-dagger operations to and from very foreign parts.

I could hardly have been expected to imagine these "special duty" operations when, in the fifth month of the "phoney war", I had looked out at those spinneyed hilltops floating on the mist and wondered whether this strange aeroplane really could be put down in the length of a cricket pitch. □

CHAPTER NINE

No Piece of Cake

*

OXFORD

VISIBILITY, under a cloudless January sky, was limited only by the hard hairline of a ground-mist horizon which was unbroken except to the north-west where subtle coruscations represented the peaks of Snowdonia.

But this was no better than a dream environment, its duration limited by daylight-time remaining and by the number of gallons in the tanks of the four Airspeed Oxford twin-engined trainers which we were ferrying to RAF Sealand from their manufacturer's Portsmouth factory. Sooner or later it would be necessary to face the realities of life at ground level.

The weather forecast had said that surface visibility might, in these temporarily stable anticyclonic conditions, drop quite sharply during the late afternoon. For one reason or another–including a delayed start because of morning fog at our Bristol base–we now found ourselves airborne at a time which could reasonably be described, on a winter's day, as "late afternoon".

So far there had been no very obvious causes for worry. Some unpleasant moments had been experienced while cruising over the western edge of an opaque carpet of fog covering the industrial areas around Birmingham–but, even then, a little comfort could still be derived from the sight of Wenlock Edge and the Wrekin, which continued to stand out in absolute clarity.

Now, cruising over Shropshire, the fog had apparently cleared below and I could see, far ahead, the silver shapes of the estuaries of the Dee and the Mersey–as well as a group of tiny points which I knew to be Liverpool's balloon barrage. And there was still nearly an hour of daylight left.

Maybe the forecast had been wrong.

Looking around for the three other Oxfords in our loose ferrying gaggle, I could see only two of them. I was not to know it then, but the pilot of the third had called it a day when he saw the fog layer over the Midlands and had put his Oxford down on that delightfully named grass aerodrome to the west of Birmingham–RAF Halfpenny Green.

How wise he turned out to be.

None of us had flown an Oxford before starting off on this trip in 1940–but we had all graduated on Blenheim bombers during brief conversion courses at the

RAF's Central Flying School and tended to look on this trainer as (to use the vernacular of the period) a "piece of cake". We were to learn more about the Oxford–and about illusive impressions of good air-to-ground slant visibility.

Not that the Oxford was vicious or even particularly difficult. It simply had a capacity for springing surprises on careless pilots. Accidentally or otherwise, this was just the characteristic needed in an advanced trainer.

The Oxford handled like a lamb when flown accurately, but could turn and give a very powerful nip to the self-respect of any pilot who relaxed at the wrong moments.

No screaming application of differential brake could, for instance, save the day if the Oxford–three-pointed, perhaps, with a bit of uncorrected drift–was permitted to deviate too far from the straight during the first few yards of its landing run. Once allowed to get out of hand it just galloped around in what is known rather absurdly as a "ground loop".

For this reason, Oxford pilots who were not instructors on the type tended to cheat by doing tail-up wheelers all the time. That was not only an unnecessary admission of defeat; it was a waste of good talent. There were few pilot-satisfactions greater than that enjoyed when an Oxford was three-pointed squarely on an out-of-wind runway and thereafter held straight without any audible loss of tyre rubber.

In many ways this trainer was unique in the era of its design and development. Whereas the basic controls of the majority of British military aircraft were, in that period, somewhat haphazard in their arrangement (or lack of it) the Oxford had a properly planned layout, with everything put where the pilot might be expected to find it.

Nor was there any nonsense about making allowances for lack of ambidexterity; the controls for its Cheetah radials, and those for the important ancillaries, were grouped logically in and around a central control pedestal. The pilot had to fly it with his left hand, using half-spectacles–and to learn to like it that way.

One little feature of this control pedestal might occasionally have caused temporary consternation for an unobservant ferry pilot who had been in too much of a hurry to wait for a cockpit briefing. Beside the undercarriage and flap controls was a little lever marked "VP AIRSCREW". This did precisely nothing–for the very good reason that the Oxford had fixed-pitch propellers. It was put there as a provision to meet a possible future need–and sensibly retained as a cockpit-drill training device.

The red disc of the sun was just touching the horizon as I eased the throttles back in preparation for a low-level inspection run over the destination aerodrome. Ahead and below was RAF Hawarden, with its monumental aircraft assembly sheds; beyond it was the estuary of the Dee; and, a little to the left, the twin aerodromes which were collectively known as RAF Sealand.

As I inched the throttles open again for a gentle shoot-up I realised that ground-level conditions were not quite as normal as they had appeared to be when seen from 5,000ft. The unpalatable fact seemed to be that there was a low-

ENGINE SPEED
BOOST
BOOST
FLAPS
DOWN
AIR
RIGHT RUDDER
LEFT RUDDER

level layer of light mist which could be seen through at moderately steep angles, but which, looked through more or less horizontally, was for all practical purposes opaque.

Pulling up in a climbing turn, I thought about possible alternative courses of action. Should I have a stab at it, or should I try to find out whether surface visibility was better somewhere else?

While I was thinking, the Oxford had already climbed up again into the dream world. At 1,000ft or so I could clearly see the aerodromes of Hawarden and Hooton Park–and even Liverpool's Speke, which was 15 miles away on the other side of the Mersey.

Obviously, there was no comfort to be gained from these visions–and the sun was hardly likely to be as accommodating for me as it had been when it stood still for Joshua. Something had to be done–and done as quickly as possible.

I could see neither of the other two remaining Oxfords; their pilots had either got on with it and landed while I was biting my nails in the circuit or they had decided to go off to some other aerodrome.

After making a second slow pass–and flying this time at approach speed with flaps and undercarriage down–I decided that things might not be quite as hopeless as they had seemed.

Even at 300–400ft it was still possible to see quite a lot. So a normal, lightly powered, fairly steep, full-flap approach could be made down to that level. There was no reason why–given a little excess speed and height for luck–the approach should not be continued successfully through a short period of near-opacity.

In the event, the arrival was successful. And that made two. The third arrival was successful only to the extent that the pilot was able to join us a little later to complain (quite characteristically) about the lukewarmness of the tea. His Oxford had ended its run in the opposite boundary fence.

It was repairable.

Not very long afterwards I was given the job of ferrying the kind of aeroplane for which the Oxford was an initial conversion trainer. A Hampden or two had already been successfully delivered, so it was supposed by the powers-that-be in the embryonically independent Air Transport Auxiliary that I would be able to manage an Armstrong Whitworth Whitley bomber.

CHAPTER TEN

Self-willed Bomber

*

WHITLEY

THE SITUATION should have been at least a mildly disturbing one–but somehow it wasn't. Maybe I imagined myself to be in such deep trouble that there was no point in worrying too much.

It was difficult enough to be doing mental arithmetic and trying to hold a reasonably accurate heading while flying over a smoke-covered countryside which had, little by little, offered fewer and fewer worthwhile landmarks against which the map-drawn track could be followed and corrected. That was a fairly acceptable state of affairs. In any case, the weather forecasts for the destination area had promised unlimited visibility under a cloudless sky and the pre-departure "actuals" had confirmed this situation.

The fact that I was conducting a Merlin-engined Whitley bomber for the first time in my life was also acceptable. That was the kind of thing which we were being paid to do in 1940 as members of the Air Transport Auxiliary civilian ferry organisation. But there was another nagging misery at the back of my mind.

Without thinking, and following the familiar routine, I'd raised the undercarriage immediately after take-off on this delivery flight from the manufacturer's field at Coventry to the RAF aerodrome at Dishforth in Yorkshire. Just as soon as undercarriage "up" had been selected I remembered that the handling notes for the Whitley had made it very clear that there was no hope, flying alone as I was, of lowering the undercarriage if, for some reason, the hydraulics failed to work.

The final emergency system for the undercarriage involved the operation–by some other, regrettably absent, crew member–of hand cranks hidden somewhere in the bowels of the Whitley's centre-section crawlway. For a moment I supposed hopefully that it might just be possible, if the worst came to the worst, to leave the aeroplane to its own devices for a while and to do the job myself. But then I remembered some of the words in the handling notes and guessed at the physical battle which would need to be waged with the machinery down below.

This didn't seem to be a particularly promising course of action. Even if the Whitley could be persuaded to carry on by itself in the reasonably smooth conditions prevailing, the lowering of the undercarriage would certainly change the fore-and-aft trim and I'd no idea how much time the operation might take. From what I remembered of the words in the book it was likely to be a laborious and long-winded business.

There was certainly nothing for me to do at that moment but to sit and sweat it out. It would be futile to select undercarriage "down" just to see if the hydraulics worked–though my right hand itched to take hold of that one among the three little levers sprouting from the floor, to lift its safety catch and to push it forward. Anyway, there was no reason to suppose that, just to spite me, the system would choose to fail on this particular flight. Why should it? Meanwhile I at least enjoyed the advantage of a normal cruising speed instead of the undercarriage-down dawdle and its prospects of excessive coolant temperatures.

So, holding the directional gyro on a compass-corrected zero, I tried to think about something else. It was more than ridiculous, anyway, that I should be in sole charge of this formidable device. My first flight in a Whitley, two years or so before, had been as a passenger, sitting in what would now be called the jump seat, during a test flight in one of the original versions. This version, with its Tiger radial engines, was the Whitley *per se*.

As my thoughts drifted back to this earlier experience I wondered whether any of those old Tiger-powered variants were still flying.

They were. A couple of months or so later two of them were to turn up on the ATA ferrying programme. They had to be taken from Gatwick Airport to the RAF maintenance unit at Kemble in the Cotswolds–and were probably among the last to be airborne. As I'd already flown a Whitley without coming to any very obvious physical harm, I was given the job of conducting one of these old warhorses on what was probably its final journey.

The Tiger-engined Whitley on my delivery chit had seemed, on superficial inspection at Gatwick, to be a fairly desperate piece of wreckage, both inside and out, but, in the event, it worked well enough. This time, though, there was to be no question of raising the undercarriage in some careless moment of absent-mindedness. This, I was told, should be retracted only if it was necessary to reduce drag in order to remain airborne after engine failure, or if some other emergency seemed likely to be about to require a forced landing.

But nobody had told me that this instruction might, even if everything went well, provide an operational problem of some slight difficulty. The Tiger engines had two-position propellers, giving a fixed fine-pitch setting for take-off and a fixed coarse-pitch setting for cruise.

It turned out in practice that neither setting was particularly well suited to normal progress with the undercarriage down. In level flight the engine revolutions exceeded the permissible maximum when the propellers were in fine pitch, but dropped more or less off the clocks when coarse pitch was selected and showed no sign of building up to any reasonable figure at an undercarriage-down cruise speed.

Then I had a bright idea. I would proceed in a series of climbs and descents, using fine pitch for the slow-speed climbs and coarse pitch for the gentle dives. This idea was successful up to a point, but the programme wasn't very satisfactory in practical terms and seemed to be giving the Tiger engines an unattractively hard time. So, after a short trial period of climbing and diving, I decided to forget the little I knew about pre-ignition and about the relationship between boost and revolutions – and settled for coarse pitch.

All went well, but when I checked in with the control people in Kemble's tower (or "watch office" as it was known then) I learned that my companion ferry pilot on the trip had taken his Whitley back to Gatwick with a report of "loss of power in both engines". I can't say that I blamed him. However, the Tigers in my Whitley–even if they had, in fact, been trying to throw off their cylinder heads while operating at revolutions for which they can never have been designed–ran like sewing machines all the way to Kemble.

The Whitley was a strange device for the uninitiated. It appeared always to be making its own very good arrangements for take-off and landing, regardless of any stubborn efforts which were applied in an attempt to make it do something else. When lightly loaded it tried to float quietly off the ground almost before the pilot had got both throttles up to take-off power. In just the same way it would drift back on to the ground again, and stay down, before it was expected to do any such thing. The only way of getting the tail down, it seemed, was to spin the trim wheel back quickly at the critical moment.

The trouble–if you could so describe this gentle self-willedness–was the result of the Whitley's basic wing fuselage design. This not only caused it to adopt an extraordinary nose-down attitude while airborne, but also upset an untutored pilot's judgment of the wing's actual angle of attack during take-off and landing.

In due course one learned to handle the Whitley with rough firmness–pushing the nose right down during take-off, so that it seemed almost to be scraping the runway–and to accept the fact that an apparently level attitude was the right one for something like a three-pointer.

One also learned, with the Merlin-engined versions, to use a lot of differential power during the first part of the take-off–leading with the port throttle and bringing up the starboard throttle to full power only when everything seemed to be comfortably "on rails".

If allowed to deviate seriously to port during take-off the Merlin Whitley demanded corrective measures which could eventually become very untidy indeed. Runways were short in those days and it was then usually better to swallow one's pride, call the whole business off and start at the beginning again.

Things were to get rather worse before they got better on that Whitley delivery flight to RAF Dishforth. Fifteen minutes or so after take-off the smog–an admirably descriptive word that had not then been invented–was absolute. Nothing could be seen down below except very occasional and unrecognisable areas of water which reflected the sunlight.

After 55 minutes I'd started to make repeated mental calculations of time and distance. These were beginning, at the end of an hour, to bring up the answer that now–or very soon after "now"–the meteorological story would prove itself to be true or false. "They" had said that the areas north of Leeds would be clear.

And so they turned out to be. Quite suddenly a whole new sunlit world of fields and rivers and roads and country towns was, almost apocalyptically, spread out ahead. More miraculously still, the city of York, with its unmistakable

Minster and its River Ouse, was lying over to the right just where it should have been in relation to the previously so-useless line drawn on my map. In a few more minutes the familiar standard-type RAF station buildings and greensward of Dishforth came into focus beyond the patchwork of mid-Yorkshire's farm country.

The undercarriage? It came down in good order. And so did the Whitley – with very little help from me.

□□□□□

Very soon, we of the ATA were to start on a new part of the learning curve – when the American twins, with their alien cockpit layouts and handling characteristics, began to appear on daily delivery lists. First for me was the Douglas DB-7 or Boston, with its new-fangled tricycle undercarriage, but only a few days later a Lockheed Hudson came up for delivery. In its own way, the Hudson presented greater difficulties for a ferry pilot who was just beginning to grow up into his extraordinary and varied occupation. □

The Merlin-engined Whitley V in the "war paint" of the 1940–41 era. Whitleys were later used as glider tugs and for troop transport, but one version, the Mk VII, remained in more active service with Christmas-tree aerial arrays on anti-submarine work for Coastal Command.

CHAPTER ELEVEN

U S Invasion

*

HUDSON

THE HUDSON'S TAIL had finally settled and the aeroplane was now, rather to my surprise, completing its landing run without any deviations from the straight. The touchdown had been, perhaps, a trifle sensational as it had lolloped along with its tail in the air, so I was glad that, in the eyes of any idle spectators on the apron, the situation would now appear to be under competent and even masterful control.

When the speed had dropped to a fast taxiing pace I eased open the starboard throttle with the idea of initiating a gentle turn in the direction of what would now be known as the "terminal area". Nothing happened; the Hudson continued to roll quietly and still fairly quickly towards the distant boundary. I grabbed the brake lever (so like that of an English Ford car of that era), pulled it through a couple of notches and applied full left rudder. Still nothing happened; the port wheel juddered but the Hudson bowled along as if on rails.

My mind worked through those sections of the Hudson's post-landing checks which still remained in my memory after the relaxation following a safe arrival. I peered at the dozen or more levers sprouting from various parts of the throttle pedestal.

Oh, yes, of course, I'd left the tailwheel lock engaged.

This was my first delivery trip with a Hudson and the little error was not the only one I'd made during a 55-minute flight from Speke, Liverpool, to the maintenance unit at Aston Down in the summer of 1940.

In terms of simple discomfort, mental and sometimes physical, these might have been described as the heroic days of ferrying by the embryonic Air Transport Auxiliary. We had long ago completed our brief conversion courses at the Central Flying School and had left the RAF ferry pools to which we had been initially attached. The ATA was now independent–but was only in the initial stages of its development. Even the score or so of picked amateurs (and near-amateurs) who had provided the nucleus of the organisation in September 1939, and were now forming the hard cores of the three civilian ferry pools, were absurdly inexperienced by any normal standards.

By means of a mixture of concentration, caution and knowledge of the country-side we got along, but only a few of the newly arrived American professional-pilot soldiers-of-fortune and temporarily seconded BOAC pilots could be considered as being capable of taking things in anything like a normal stride. For the rest of

us, every new aircraft type provided a strange and sometimes mildly harrowing adventure. We fought our way through half-remembered check lists while worrying about the weather reports and about the daylight time remaining.

When the weather was good and ferrying programmes could, with luck, be guaranteed, life was pleasant enough. We enjoyed the work and some of us had leased flats or houses in which we could forget the days' unpleasantnesses and squalors. But during bad-weather periods we tended to feel ourselves to be homeless and shabby nomads. Our uniforms became grubby and were stained with hydraulic fluid; we bought shirts and underclothes to supplement the meagre contents of our over-night bags at the different points on our devious delivery travels.

Once, during the not very pleasant spring of 1940, two of us spent nine days flying a pair of worn-out Wellingtons from Marham in Norfolk to Aldergrove in Northern Ireland. Yet–such are the vagaries of the English weather–we managed, on the ninth day, not only to take the Wellingtons on the final sector from West Freugh to Aldergrove, but to motor over to Queen's Island, Belfast, collect and deliver two Handley Page Hereford bombers to an aerodrome in the Cotswolds, and be picked up there and flown back to our base before dusk.

□□□□□

The Hudson was, for me, very much of an off-beat aeroplane; the sight of the complications of its cockpit caused an intensification of the familiar contractions of the stomach. I had learned to find my way with comparative ease around the

disorganised cockpits of British aeroplanes, but, as I settled into the pilot's seat of this American device, I wondered seriously, for the first time in ten months of ferrying, whether my bluff had finally been called.

Even the engine-starting procedure seemed to be far beyond the capacity of two hands and their ten fingers, but eventually–after working with the cunning of a cinema organist at the starter and booster-coil buttons, the engine selector and doper switches, and the wobble-pump–the Wright Cyclone engines were encouraged, one after another, to burst into life. As they warmed up and I looked around the cockpit, my renewed confidence was already evaporating.

How on earth, for instance, could anyone learn to control the fore-and-aft trim with that little crank at the back of the throttle pedestal? I wound it this way and that, watching the indicator dial. It seemed that if I turned it clockwise the nose should go up, so I would think of it that way: screw in, nose up; screw out, nose down.

Everything went fairly well on the take-off–though, forgetting for a moment that there was no automatic boost control, as on British aeroplanes, I allowed the limitations to be exceeded for a second or two before bringing the throttles back.

Fortunately, the weather was near-perfect and there were no navigation worries. Once levelled out on the cruise, maximum attention could be paid to the curious and involved machinery around me. I messed about with the fuel system; I reduced power and speed, practised with the trim crank and experimented with the effects of various flap settings; I watched the exhaust-gas analysers while playing with the mixture controls; and I set the wing de-icer boots off on their slow pulsations just to see how they looked when in action.

Finally I adjusted the instruments on the centre panel and cautiously moved the autopilot engagement lever. Nothing whatever happened: left to itself the Hudson made a slow turn to port and dropped its nose. Perhaps, I thought, the autopilot needed a moment or two in which to warm up. Then I noticed that the flap lever was still in the "up" position and moved it to neutral.

Immediately a giant took charge of the controls, settled the Hudson firmly back on to its original heading and levelled it out from a gentle dive. The autopilot obviously *did* work, once it had been allowed some of the hydraulic pressure which had been vainly exhausting itself in the flap circuit. At that moment the

starboard Cyclone engine chose to frighten me with a resounding hiccough. I pulled the mixture levers back. That, I thought, would teach me not to fool about with things I didn't really understand.

The familiar aerodrome known as Aston Down (often confused with nearby Kemble) was now in full view over the nose. The awful moment had arrived; I had to put the Hudson safely down on that little green patch.

While my right hand fluttered over the various levers on the throttle pedestal (a baker's dozen of them, in fact, not counting those for the undercarriage and flaps) I tried to remember all that I'd been told, rightly or wrongly, about this formidable aeroplane. Never use full flap; never trim it right back on the approach; motor it in to a wheeler; and don't try to three-point it. These and other things (which escaped my memory) they had said.

As the grass came up, instinct forced me to haul back on the control column in a valiant effort to make a "proper" landing. The main wheels touched rather heavily and I knew then that it was hopeless to try to play it out in the old familiar way. I pushed the control column hard towards the dashboard and waited. After an agonising second or two the wheels touched again and then again –and stayed on the ground. Not my idea of a landing, but it seemed that the Hudson had, in its own way, duly arrived.

□□□□□

Meanwhile, we had been grinding out scores of hours on an aircraft the flying of which was becoming second nature–the Avro Anson. Once we had begun to retire some of the more ineffective oddments which were used initially the Anson became the ATA's normal taxi vehicle. At first it was flown by odd-men-out on what would now be called positioning flights, and by the most experienced or senior pilot when there were ferry-pilot passengers aboard. Later, the ATA appointed specifically named pilots to handle the Ansons throughout their long and complicated taxi rounds. The story which follows is built mainly around one of the comparatively rare occasions when an Anson was actually being ferried on a delivery flight. □

CHAPTER TWELVE

Faithful Annie

*

ANSON

ACCORDING TO THE FORECAST the cold front should have been through by midday, but at 1300hrs the cloudbase was still a solid scudding greyness at 500ft or so and the rain swept in curtains across the aprons at Woodford aerodrome.

I had been expecting to be able to take off before noon with a brand-new Avro Anson for delivery to the RAF's No 23 Maintenance Unit at Aldergrove in Northern Ireland. If I delayed departure much longer the taxi pilot waiting there would have to give me up and set off on the remainder of his round of ferry-pilot deliveries and collections.

It was a day in June 1940, and there were many hours of daylight–but the taxi pilot also had many more jobs to do. I knew all about them; I had worked out the programme for No 3 Ferry Pilot's Pool at Hawarden before being dropped off at Woodford in advance of the expected front. He had been stuck at Belfast the previous evening and would not relish another uncomfortable night away from base.

Nor would I. The latest "actuals" showed that Jurby, in the Isle of Man, must have been enjoying post-frontal clear skies for at least two hours. Speke, Liverpool, was still reporting a 700ft cloudbase and continuous rain–but the clearance ought, I thought, to be reaching the coast of Lancashire very soon now.

Luckily the factory ground crew was as keen to see the Anson on its way and off their hands as I was to get started on this delivery flight. They cheerfully put on their oilskins to do laborious battle with the Cheetah engines' handle starters and to face, with their backs to the driving rain, the dreary business of engine run-ups and chock removal.

Once I was airborne, it became obvious enough that there was no very generous amount of airspace between the ground and the tattered cloudbase under which the rain reduced visibility to five miles or less. So as to be free later to concentrate on the business of keeping clear of balloon barrages by map-reading and "Bradshawing" my way from Woodford to the coast, I made three wide circuits while winding up the undercarriage.

Even with well-used Ansons this was no mean chore. The collapsible handle on the right of the base of the pilot's seat had to be cranked through an incredibly large number of turns before the undercarriage was up–and again, later, before the red light on the dashboard went out and two little green knobs popped up beside the throttle pedestal to register "undercarriage down".

With a factory-new Anson the mechanism was always very stiff. The operation was one for a trained oarsman rather than a pilot – and certainly not for a pilot who needed, as I did on that day, to look where he was going.

Until the hoped-for weather clearance came my plan was to fly a compass-cum-gyro course to the Manchester Ship Canal. Then, when this unmistakable landmark and the Manchester–Liverpool railway had been crossed, I would map-read my way through the industrial complexes around Wigan and on to the line for Southport. This seemed to be the best way to play it. The balloons over Manchester and Liverpool were up and there is high ground to the north of Wigan.

All went well until I decided to edge the Anson around a more than usually heavy curtain of smoke-impregnated rain and to make a short cut to the south of Wigan. One railway line and then another went by underneath. The next, according to my map, should be the one which would lead me to Southport and to the Irish Sea with its comfortingly flat and low-level surface.

I turned on to it, thankful for the Anson's bad-weather windows, and relaxed a little.

But not for more than about 30 seconds. The railway line suddenly disappeared

An Avro 652*A Anson GR Mk I.*

into a tunnel–and there was no such tunnel marked on my map. I glanced down at the compass and turned the verge ring so as to line up the grid wires with the needle. The heading was 260 degrees. It should have been about 300 degrees. We were following the wrong railway line and the Anson was heading straight for Liverpool's balloon barrage.

Long before the shock of the discovery could make any consciously physical impact, a flush of adrenalin had–like the blink of an eyelid before a fly has been consciously seen–banked the Anson over to starboard. The township Ormskirk –its buildings recognised from childhood memories–rushed past on the port side.

□□□□□

Myths are built up around nearly every aeroplane. For the Anson the myth was one of reliability and harmlessness. Annie, the Old Lady of Woodford, could never, never turn and bite you; her Cheetah engines never stopped; and any fool could handle her.

In fact, the Anson could, like any other aeroplane ever built, turn and bite; the

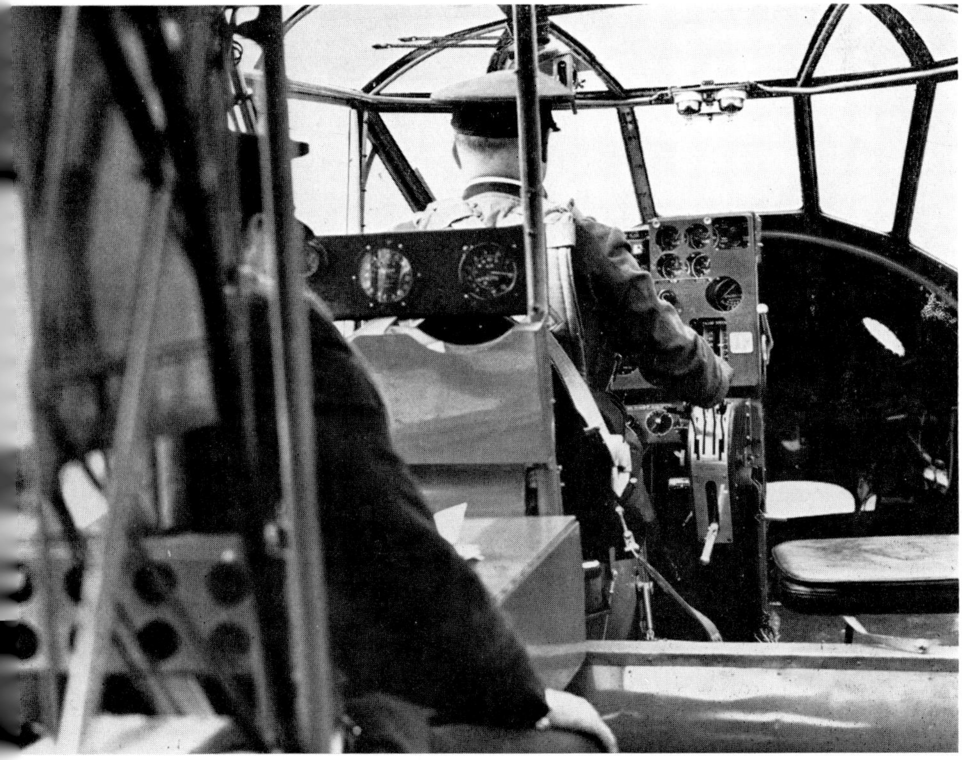

The flight-crew stations of an Anson Mk I of 1938.

Cheetahs did sometimes (though very rarely) stop before the cut-out wire had been pulled; and it was not particularly easy to fly. But it was a forgiving old thing.

The interminable windings for undercarriage retraction were an acceptable agony and the work kept pilots and pseudo-crews in better physical condition than they might otherwise have enjoyed. The sliding fuel cocks to the right of the second pilot's seat will be forgiven even by those who lost so much skin when operating them from the driver's seat on the other side of the aisle. Ground (and some flight) crews may even remember, with no more vicious reaction than a nostalgic shrug of the shoulders, their struggles with the engine-starting handles.

Those of us who lived with the Anson as a taxi hack from day to day will not forget, either, the desperate feeling of helpless frustration when the brake-operating pressure in the ground-filled air bottles began to fail when we were turning on to the runway for a last, late evening trip. And if the throttles were carelessly handled at the beginning of the take-off the Anson could, like any other twin, dash around in a humiliatingly uncontrollable semi-circle.

The "book" said that it would not remain continuously airborne when flying on one engine–but a test pilot friend of mine flew one successfully on the starboard engine from somewhere north of Worcester to RAF Aston Down on the top of the Cotswolds. When safely overhead this aerodrome he gained as much height as he could, using override power on the good engine. Then he spun the undercarriage crank through its scores of turns and made a conventional pre-1940 dead-stick approach–complete with S-turns, sideslips and all the rest of it. He had been brought up in the era when all landings were forced landings and in an environment where it was nothing less than shameful to drag an aeroplane in on the engine(s) at tree-top height.

But, myths or not–and physically exhausting though its operation could sometimes be–the Anson was, statistically, a remarkably safe device. During several hundreds of hours in its pilot's seat (and many others spent sitting on parachute packs on its fuselage floor) I experienced only one unpleasantness which was not due to simple misjudgment or sheer idiocy.

During what would now be known as a Check 4 at my maintenance unit, the fore and aft trim controls had been accidentally reversed. After the effects of the initial shock had worn off during the early stages of the test flight, trimming in reverse proved to be rather less difficult than driving a motor car to right-of-the-road rules.

For some of us the Anson had special memories. It was the aeroplane into which the ferry pilot tumbled thankfully after the last delivery flight during a sometimes harrowing day. It was, for many RAF maintenance units, the communications hack; a test pilot at one of these units breathed a sigh of relief when the day's programme required him to do no more, perhaps, than to make, with the Anson, a round of equipment-collecting visits to other units and factories.

Of course, the Anson was sometimes treated too carelessly by people who had not yet learned that all aeroplanes are, by the nature of the laws of Sir Isaac Newton, intrinsically dangerous.

Once, while making a final approach in an Anson, I noticed some surprising and variable changes in the fore-and-aft trim. The landing was completed without incident, but I was surprised that my test-pilot passengers were so insistent that there would be no need to put in a report to the unit's chief technical officer. It transpired that these over-healthy young men–on a test-flying "rest" between operational tours–had been "playing harps" with the bare trim wires running through the fuselage.

For me, there is one reason why the Anson is an exclusive aeroplane. In it I learned to fly with my left hand while doing all kinds of other things with the one which had been originally taught most of the tricks.

□□□□□

The sky was lightening ahead as, in a panic, I had banked the Anson away from Liverpool's weather-hidden balloon barrage.

In less than a minute we had flown out of the miseries into the post-frontal paradise of broken cloud, blue skies and near-unlimited visibility. As the Anson was climbing up across the coast of Lancashire I could already see the clouds which still topped the Isle of Man's Snaefell.

There was surely no pleasure much greater for the pilot of earlier days than the first sight of the bright lines of an approaching weather clearance.

□□□□□

A few months afterwards I succeeded in getting into the RAF and was duly posted, a very amateur officer in the general duties branch, to No 48 Maintenance Unit, Hawarden, as a test pilot. In those days, with bombing raids expected nightly, all eggs were never distributed in one basket and these units held the most varied assortment of aircraft. At Hawarden they consisted primarily of Hawker Hurricanes (see chapter 6) and Vickers Armstrong Wellington bombers (see chapter 14) but a lot of other strange devices were tested. Among them was the Bell Airacobra. □

CHAPTER THIRTEEN

Back-to-front Fighter

*

AIRACOBRA

TWO DAYS before Pearl Harbour, in the third December of the War, the Ministry of Aircraft Production organised a little display of some of the newly delivered American military aircraft. This hands-across-the-sea party was held at Speke Airport, Liverpool, where so many US aircraft had been assembled and modified.

Not all the new (or newish) American types were available on the spot and a missing one was the Bell Airacobra fighter. The nearest examples were at the RAF's No 48 Maintenance Unit at Hawarden near Chester and an imperative signal came through to this unit from the Ministry via 41 Group headquarters. An Airacobra was duly made ready and I volunteered to take it over to join the display. What could be nicer than a half-day off from a test-flying routine – coupled, maybe, with a good official luncheon – for Pilot Officer Prune ?

The Airacobra was, with its single rear-mounted engine, and tricycle undercarriage, quite a centre of interest at Speke. But no one, until I arrived there, had told me that I would be expected to give a flying display. Air Traffic Control gave me my take-off time and explained that there would then be a warm front going through, that there might be continuous drizzle, but that the cloud-base would not be unreasonably low.

In due course I taxied out for take-off, glad that nobody could now expect much in the way of an aerobatic exhibition – which, since my air experience with the Airacobra then so limited, could only have been personally disconcerting as well as indifferently executed.

Almost as soon as we were off the ground the Airacobra was flying through the ragged streamers of cloud trailing below a more solid base. So much for the forcast. I made some kind of rough and ready circuit on the little that could be seen of the local landscape, dived past the VIPs standing on the apron and called the demonstration off.

Not that, without radio contact, I could tell anyone about this decision – but, by way of reciprocal advantage, no one could tell me what I ought to try to do in the way of a demonstration. There was, in any case, one of the new North American Mustang fighters somewhere in the circuit; I had seen this Anglo-American fighter preparing to take off behind me and I reckoned that a couple of aeroplanes would be just one too many in that part of the sky.

There is a curious, if ephemeral, feeling of security in the simple fact of being airborne – even in the worst of weather. A decision has to be made, but not,

"The cockpit was like the inside of a miniature single-seat saloon car . . .".

maybe, during the next half-minute. Seconds become minutes in such circumstances and one is free to ruminate about one thing and another.

The Airacobra–or P-39, to use the official designation of those days when American fighters were "P for pursuit ships"–was an extremely ingenious, but strange and, in some ways, rather nasty little aeroplane. Its Allison liquid-cooled Vee-12 engine was mounted behind the cramped cockpit, ready to plunge itself into the pilot's kidneys, and drove its propeller through a shaft running between his outstretched legs. This shaft whipped and rattled noisily while taxiing and we soon learned that the only way of dealing comfortably with the Airacobra on the ground was to hold constant and rather high power and to drive it along fast, using the pedal brakes to hold it in check.

The cockpit was like the inside of a miniature single-seat saloon car which one entered through a little door on the starboard side. The view through the heavily sloping, curved windscreen was just about acceptable. When the pilot had finally tucked himself into it, the interior situation could only be described by the word claustrophobic.

But the Airacobra was a beautifully engineered device. By the end of the second year of war, test and ferry pilots had become accustomed to tricycle undercarriages and to the confusing tidiness of American control cabins–the layouts of which made it so obvious that everything had its proper place.

The barbarous English preferred an apparently unplanned chaos in their

UF
J
L

43572

cockpits, which were just about as tidy, in that era, as village blacksmiths' shops. Somehow things seemed to be less complicated that way and, if one searched around for excuses, it might have been argued that big and important jobs, such as raising the undercarriage or lowering the flaps, should be done by big and important-looking levers which were isolated from all the other controls.

In the Airacobra–and, for that matter, in much bigger aeroplanes such as the B-17 Fortress–the undercarriage was raised and lowered by means of a tiny little tumbler switch. This was very neat and rational–since the actual job was being done by an electric actuator–but it didn't seem quite right. Amidst all this electrickery and rows of switches, however, one usually found that the really final and vital things were done, even in American aeroplanes, by basic, mechanical blacksmith methods. In the Airacobra the emergency undercarriage-lowering system was operated by a hand-crank on the right of the pilot's seat.

Meanwhile, here I was, airborne in this cramped, unfamiliar and complicated little aeroplane and something had to be done. We were flying over the brown waters of the Mersey Estuary at near-zero altitude and with nothing much to be seen through the drizzle-streaked windscreen.

I could neither go back to Speke nor climb up through the cloud in the hope of reaching the clear. For one thing, I had no idea in which compass direction any possible "clear" might be found; for another, the first rule to be obeyed by those in aeroplanes without radio is to stay well below cloud. It is, unfortunately, impracticable to attempt to stay in one place. All I could do was to fly in the general direction of home and to be prepared to think again when the south bank of the Mersey appeared through the murk.

Then I remembered Hooton Park aerodrome–which was not very big, but big enough, and should be lying more or less where the nose of the Airacobra was pointing. Just as soon as land appeared I turned and aimed at the place where I thought this aerodrome ought to be, selected flaps and undercarriage down (how glad I was of the little switches), closed the throttle and prepared to land. Had not Winston Churchill said, a long time ago, that aviation must fly by itself? That was exactly what aviation would now have to do for me.

A hedge went by underneath. It might be the boundary of the aerodrome, or it might be the boundary of the field just short of it. The undercarriage-locked-down indication came up on the dashboard as the Airacobra touched. We bucketed along–wheels skidding on the wet grass as the brakes were uselessly treadled–and something which looked rather like an embryonic building estate loomed up through the drizzling gloom ahead.

This was, in fact, Hooton Park aerodrome, but I had entirely forgotten that a runway was being laid down there. Everyone was very polite and eventually the Airacobra was manhandled out of the artificial ditch in which it had ended its landing run, trundled past a concrete mixer and a row of drainpipes and parked on a bit of open turf. In due course the weather cleared, and I started up, taxied out, chose the longest run to be seen on what remained of the grass aerodrome and took off for home.

A brace of Hurricanes and a Wellington were waiting for test when I landed at Hawarden ten minutes later. Nobody asked me whether I had enjoyed myself or whether I had had a good luncheon. And I was quite satisfied just to be around the place and still in one piece.

□□□□□

But this was a mere interlude; Hurricanes and Wellingtons were the aircraft with which we were primarily concerned at No 48 Maintenance Unit. □

CHAPTER FOURTEEN

Geodetic Variations

*

WELLINGTON

THE SUPERCHARGER of the port engine of the Mark III Wellington bomber had disintegrated, but there had been nothing particularly dramatic about the event. Only on the ground afterwards, when the cowlings had been stripped off, was it possible to appreciate the extent of the damage caused by the bits from the burst impeller. Nearly every intake elbow had been split open and the Bristol Hercules powerplant looked as if it had suffered a hit from a cannon shell.

In the air the effects had been comparatively innocuous. As I checked the operation of the supercharger gear-change there had been a more than usually vigorous kick from the port engine and I found that I was pressing rather hard on the starboard rudder pedal. Revolutions and boost were temporarily unchanged and a few seconds went by before I realised that this engine was no longer giving any power. When I looked at it I saw that a lot of jagged holes had appeared in the cowlings and that these holes were emitting little jets of blue smoke.

We were about 20 miles from the home airfield and a group of Air Training Corps cadets were flying with me on an air-experience outing. The port engine drove the pump which supplied the normal hydraulic pressure. If I feathered its propeller the undercarriage and flaps would have to be lowered by hand pump, and I knew how long that would take with no one to help.

Not a very clever situation. I looked again at the dead engine, Smoke was no longer coming out of the holes and there were no unusual vibrations as its propeller windmilled round. All right; I would break the rules and feather the propeller only after the hydraulic pump had done its job of lowering the undercarriage.

As we entered the circuit I selected undercarriage "down", watched with relief as the green indicator lights came on, and pressed the port feathering button. The propeller slowed down and stopped with a jerk.

All I had to do now was to line-up towards the runway-in-use, come in with height to spare and adjust the approach on the flaps by working away at the hand pump. The Wellington floated in over the threshold a bit high and a bit fast, and one of the cadets chose that moment to ask: "Why has the flight been so short sir?". But all was well and I even had time to choose a suitable parking space as we turned off the runway.

Whilst waiting for the maintenance unit's ground crew to find some transport and to motor round the perimeter track I wondered how people mastered these

quite ordinary but complicated little tricks in the judgment and execution of approaches and landings. I thought back to the days, a dozen or more years earlier, when I'd been learning to fly.

I was certainly no prize pupil. During that summer of 1929 I'd floundered about the sky in a de Havilland Cirrus Moth for the better part of a dozen training hours before going solo–and had wondered again and again whether I would ever manage to learn the tricks.

During the first hour or two in the air I'd been conscious only of an horizon at which some selected part of the front end of the aeroplane had to be aimed. With this Cirrus I Moth there were the clacking, open valve rockers to distract attention in front and a hot exhaust pipe to burn my cheek if I leaned my head too far out to port.

Those were the realities; for the rest, the Moth might just as well have been hanging on a skyhook over an enormous grey-green bowl. I'd sat on the floor of its narrow fuselage–inexplicably buffeted by a full gale from which the tiny plastic screen gave little protection–and had tried, by means of a succession of interrelated and imperceptible pressures on the rudder pedals and control column, to make this fuselage keep still, tilt down, tilt up, or swing round along the hypnotic horizon.

After a while the control movements had begun to make some sense and I eventually learned, without the exercise of unreasonable concentration, to make the thing move about within the bowl more or less as the instructor required it to be moved.

Then, just as these most elementary reflexes had been trained and some relaxa-

TOP LEFT: *Inside a rare version of the Wellington Mk I with primitive dual control.* ABOVE: *A Wellington III.* RIGHT: *First and last Wellington – the prototype K4049 in July 1936 and a T Mk X, MF628 photographed in 1953.*

K4049

F628

tion was beginning to be possible, the instructor had swept me out of the dream environment and down to the bottom of the bowl.

It was all very well, it seemed, to have learned how to control a vehicle which apparently hung stationary in the sky and just rocked or revolved around some imaginary point. The same control movements had now to be made with large and very solid objects rushing by, and these moving objects, of constantly changing shape, had to be slowed down at the right moment and at the right place so that an arrival could be arranged.

No longer could I lean back and recover lost sang-froid while the instructor miraculously found his way back to the aerodrome, and even more miraculously put the Moth neatly on to that little patch of open ground. I had to be taught to approach and to land.

Theoretically, the trick of landing an aeroplane is as simple as the instructor's "patter" which, in those days, accompanied the operation. "The aircraft", the instructor said (trying, at the same time, to make the events fit the words), "is held off the ground by a steady backward movement of the control column until all flying speed has been lost. It is then landed, undercarriage and tailskid together."

Which was fine for the supernaturally skilled; I used the greater part of my limited experience and skill in devising ways and means of dealing with the situation when the undercarriage and tailskid had somehow *not* managed to touch the ground at the same moment, or anything like it. Then the aeroplane leaped wildly about and one of us (usually the instructor) had to calm it down with bursts of throttle and wagglings of the control column.

The entire operation had seemed to be bordering on the impossible until the moment when my clumsy, carefully thought-out control movements had suddenly and unexpectedly begun to turn into natural reflexes. At the same time, memory had been unconsciously registering the changing shapes and sizes of houses, boundary fences, hangars and the familiar grass aerodrome as an approach is being made from the right height and at the right speed.

Somehow a pupil knows quite well when he is due to go solo. So I was ready for the stomach-contracting shock when the instructor climbed out on to the wing, reached for his cushion, and slowly and laboriously arranged the lap-straps in the front cockpit so that they wouldn't get in the way of the controls. "Off you go," he said, "if you feel happy about it. Just one circuit and taxi her in."

Giving myself (and my stomach) no time to think about the prospect, I glanced up to see that no aircraft was on its approach, moved the "cheese-cutter" fore-and-aft-trim control to neutral, turned into wind, rolled for a yard or two and opened the throttle. I forgot that the Moth now had only one body to carry and was surprised and a little alarmed when it became airborne so soon.

The situation had all the characteristics of a dream. It was a quiet July evening, but the natural brightness had been toned down by industrial smoke. This had given the familiar countryside, the local power station and the nearby tyre factory a transparent and colourless appearance.

I was cold and lonely, but feeling fatalistic rather than frightened. I missed the familiar helmet in front and was not comforted by the sight of the instructor's tiny figure as he walked back across the aerodrome to the clubhouse, the white dot of his face flashing occasionally as he looked up to see where I'd got to in my circuit.

For the rest, I had been, on this first solo, almost an automaton, faithfully following, so far as skill permitted, the patterns of action and judgment with which, to complete the analogy, I had been "programmed" during the earlier circuits.

It was not, in fact, a very good first solo. I landed rather too far up the field and had to apply a poorly judged burst of throttle to control a bounce before the Moth finally settled. But a first solo flight is, after all, no more than a symbol of elementary proficiency. The real learning starts from that moment.

While the Wellington, with its port engine feathered and its undercarriage locked down, had been floating in on that semi-forced landing, the changing visual pattern of the runway and the "feel" of the aeroplane had been instinctively and continuously compared with the memories of two thousand other flights. This approach was just a bit more important than most of the others had been. It had to be good; there'd be no second chance.

Even by the individualistic standards of its day, the Vickers Wellington was a very strange aeroplane. Not only because, like its single-engined predecessor, the Wellesley, it had a unique form of construction which was effectively but not quite accurately described as "geodetic"–usually spelt with a capital "G". In nearly every other way the Wellington was special. It was almost as if its designers had worked out all the details for themselves on a desert island, using, perhaps, outline drawings of other advanced aeroplanes for guidance, and importing, ready-made, only the engines, propellers, instruments, wheels and tyres. Nearly every other part of the Wellington was uniquely conceived, designed, shaped and manufactured.

None of which is meant as a criticism of the Wellington. Once a pilot had become accustomed to its minor quirks of behaviour and its oddly assorted controls, he found it to be a delightfully easy and forgiving affair. However much the structure quaked in rough air–with the control column wandering erratically backwards and forwards as the elevator cables moved to match the flexings of the fuselage–no one ever had any doubt about the strength of geodetic construction. We knew, too, that it was *meant* to flex. No one took any serious notice of the pilot's jokes about the lazy-tongs principles and the impossibility of measuring the Wellington's span or length except at some instantaneous moment. Had we been knowledgeable enough in those days we might have made some quip, with academic confidence, about this practical application of Heisenberg's uncertainty principle in nuclear physics.

I had had only one fairly brief ferrying ride, sitting beside the pilot on a much earlier trip, before I lifted myself (I hoped) athletically up and through that shin-barking cockpit trap-door to take a Wellington IA from Brooklands, where it had been built, to RAF Feltwell, in the flat, cultivated wilds of Norfolk during the first April of the war. While clambering up into that dais-like pilot's seat on the left, I was already running my eyes over the control arrangements.

There was the unusual rounded-rectangular-shaped control half-wheel, with its matching pair of wheel-brake levers which were operated by squeezing with all the fingers of one or both hands. At the base of the centre panel was the undercarriage control with its first-and-second-finger-operated safety catch. Beside it was the flap control. Later, this simple little up-down or neutral control was to be fitted with a "leak" device to slow up the rate of retraction on overshoot, and with an interconnection arrangement for the fore-and-aft control to compensate for the considerable changes in trim when the flaps were raised and lowered. In 1940 we had neither of those aids to less troublesome and tidier flying.

The elevator trimmer on the pilot's left was a spoked affair, called an "asterisk" –rather than the more usual wheel. The engine-cooling gills were opened and closed by means of a crank below and to the right of the pilot's instruments. Just about the only familiar cockpit feature was the standard RAF six-dial blind-flying panel.

In those days no one knew, or needed to know, anything about rotation and lift-off speeds–though we began to learn about them later with the B-24 Liberator, and those of us who had been through a conversion course at the RAF Central Flying School knew about minimum engine-failure safety speeds. We just raised the tail and waited until the aeroplane felt as if it might be ready to fly. At Brooklands the rim of the racing track formed a hard and obvious obstacle which just had to be cleared, but it was necessary to resist the temptation to try to make the Wellington fly before it was good and ready to do so. This lightly loaded Mark IA was, in fact, comfortably airborne long before the so-hard edge of the top of the track had begun to look like a menacing reality.

The air was a bit rough during the climb to cruising height out of Brooklands and it was then that I was aware for the first time of the curious elevator-control wanderings which were to become one of the accepted characteristics of the Wellington. The column just moved gently and erratically fore and aft except when the air was reasonably calm.

It was the acceptance of this feature which led to difficulties when trying, a long time afterwards, to sort out some extremely queer fore-and-aft behaviour in one particular Wellington. I had just returned to an RAF maintenance unit from a fortnight's attachment to a Vickers assembly plant where I had been testing production-line Wellingtons; so I thought I knew something about standards of handling behaviour. Two days before I had got back to the unit this particular IA had demonstrated some abnormal fore-and-aft control characteristics about which nobody–test pilots or engineer officers–was prepared to be very specific. That the Wellington's elevator control and fore-and-aft trim behaved in a very odd way were just about the limit of the factual matter in reports from the unit's test pilots.

As a temporary "expert" I was expected to offer a more definitive judgment. Accompanied by the flight sergeant in charge of the "site" which had originally prepared the Wellington for test, and trying to ooze confidence, I taxied it out and took off. Even before we were airborne it was obvious that there was something more than normally odd about this Wellington. It was erratically unstable and seemed to be trying to do things on its own a second or two before any control action was taken. The behaviour just did not conform to any known pattern, and my "expert" judgment was no more specific than anyone else's.

The rogue Wellington sat for two more days at dispersal while the senior engineer officer considered the matter. Then, bored with inevitable inaction, he drove over to dispersal and took a walk around it. Irritated, no doubt, by its monstrous and silent inadequacy, he struck the port elevator with the glove which he was carrying in full formal "site"-visiting rig. The elevator made an unusual rattling sound. Riggers were called into action and part of the fabric was cut away for his inspection. A section of the structure was missing, and part of a kit of tools was laid out in full view. Not only had the fabric been flapping about in the structurally unsupported area, but the tools had been wandering around inside the elevator during at least one ferry delivery and during several abortive test flights since the Wellington had arrived at our unit.

Clearly, it seemed, one shift had left some work half done and the next, assuming that it had been completed, had just sewn up the fabric. I cannot remember whether anyone tried very hard to discover the identity of the culprits; probably not. The Wellington had been in squadron service and at one other maintenance unit before it reached ours, where no work of any kind, except a daily inspection, had been done. I still wonder what the Air Transport Auxiliary pilot thought about the handling on its delivery flight and why he (or she) had not reported that there was something queer about it. Maybe this had been the pilot's first ferrying trip with a Wellington and the erratic fore-and-aft instability had been taken for granted as being typical of its hearsay behaviour.

Wellingtons came in more than a dozen forms–from the original Bristol Pegasus-engined I and IA to the various Coastal Command variants which carried mines and were sometimes graced with a submarine-hunting ASV radar aerial system of drag-inducing complexity. In between were the bomber Mark II with liquid-cooled Rolls-Royce Merlin engines, the III (and later the X) with Bristol Hercules radials and the IV with Pratt & Whitney Twin Wasps. Strangest of all were the Mark V and VI, which were experimental, pressurised, high-altitude bombers fitted with Hercules and Merlin engines respectively. Only three of the Vs were built, but more than 60 of the VIs were turned out–and used for various purposes other than those for which the variant had been designed.

In 1942 we had two Wellington VIs at a maintenance unit to which I was attached. For the flight crew this aeroplane, with its air-tight capsule and submarine-like entry door was a highly claustrophobic affair. The crew were, in effect, locked in a kind of pressure cooker, the temperature of which was made even more apparent for the pilot because he sat with his head inside an astrodome through which the sun could shine mercilessly. It would probably have been even more claustrophobic for the occupant of the tail turret if the Wellington VI had ever been used in hot blood; he had a little pressure cabin all to himself.

All Wellingtons were somewhat strange devices–and this Mark VI was the epitome of strangeness–but we liked them because they were easy to fly and because we knew they were tough old things. Even when, on one routine test, the fabric started to tear away from the port wing, there seemed to be no particular cause for alarm. Perhaps I believed that the Wellington would continue to fly on

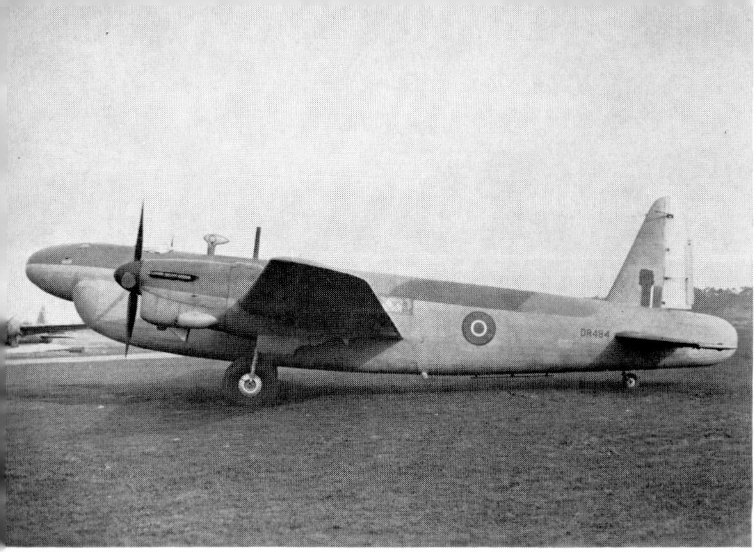

ABOVE: *Wellington Mk VI.* ABOVE RIGHT: *Wellingtons under construction in* 1939 *showing the geodetic construction of the fuselage.*

its geodetics alone–or maybe I'd been over-persuaded by that much-published photograph of one which had returned from a raid with all the fabric missing from the fuselage aft of the wing. It must have been very cold and draughty in that Wellington during its long flight back to base.

The principle of geodetic construction may now be considered to have been a "sport"–to use a biologist's word–in aircraft construction methods. Stressed-skin development had already superseded it in the middle 1930s. But let us not forget that the single-engined Wellesley, first of the geodetic aeroplanes, broke the long-distance record in November 1938. Two of a flight of three Wellesleys of the RAF's Long Range Development Unit flew from Ismailia (Egypt) to Darwin (Australia) non-stop–a distance of 7,162 statute miles. The third had to land for fuel at Koepang (then Dutch Timor) after flying 6,600 miles.

That was more than 30 years ago. The fuel load needed to cover such a distance could not have been lifted (with only about 1,000hp) and carried economically by anything but an extremely efficient aeroplane. Yet even in 1939, when I flew a Wellesley for the first and only time, the type was already out of squadron service. I hate to think of the power, weight and runway requirements of the B-52H which later held the record of very much less than twice the distance covered by the Wellesleys in those long-ago days. *Sic transit* . . .

□□□□□

While clattering around in our conventional (or unconventional) British aircraft we were, at intervals, learning to do battle with American heavies–the Boeing Fortress and the B-24 Liberator. The B-24 was the first of the "modern" aeroplanes with high wing loadings and involved cockpit checks. □

CHAPTER FIFTEEN

Prototype for Professionals

*

LIBERATOR

THE THRESHOLD of the short runway at Prestwick was coming up on final approach so I started to sing out the indicated air speeds from the second pilot's seat. "One-thirty . . . one-twenty-eight . . . one-twenty-five . . . one-twenty-three . . . one-twenty-three . . . one-twenty and a bit . . . one-twenty . . . hold it . . ." The control column came slowly back towards my chest and the Liberator hit the ground with the usual rattling jolt.

"Throttles closed," ordered the first pilot as the nosewheel settled. He started to brake with the approved intermittent applications of toe-pedal pressure. The applications were, maybe, a bit more prolonged than the handling notes recommended–but then we had touched-down rather on the fast side and were already a third of the way along this little 4,500ft runway. "Flaps up . . . gills open . . . cut outers," the first pilot added in conversational tones as the Liberator was turned, tyres squealing, on to the perimeter track.

This was the last day of a week's conversion course on the American B-24. After eight or so hours of dual, the two pupils on this course had been let loose on it for two days of flying practice, taking our turns in the first and second pilots' seats and pretending to be captain, first officer, or flight engineer as the occasion demanded. For personal interest I had even done a turn as first pilot in the right-hand seat–peering out through the opposite screens as we made our left-hand circuits and blaspheming in general terms about the approach situation.

For me the exercise was by way of being something of a busman's holiday. I had already put in about ten hours of test-work on Liberators at a maintenance unit, where the chief test pilot and I had taught ourselves to fly this mammoth after a brief series of demonstration and instructional circuits with a visiting 41 Group pilot. So I could relax a little and try to enjoy the new environment–and a little of the freedom of wartime hotel life by the seaside–while my co-pupil sweated in his spare time through our one copy of the handling notes.

The Liberator III was a complicated device by the standards of 1942 and behaved like an amply-powered cartload of bricks. The aeroplanes to which we had become accustomed could still be 'floated," power-off, on to a runway; this B-24 had to be motored until it touched, and the rate of descent had to be controlled by carefully judged applications of power. Nor did it just fly off by itself; the Liberator had to be lifted off when the airspeed indicator showed such-and-such a figure–and not before. If lifted too soon at high weights it was inclined to

"get stuck", uncomfortably gaining neither speed nor height. And, cruising at these higher weights, we learned that all bumps were down bumps.

The complications of the later B-24s included turbo-blowers, fuel booster pumps, an hydraulic booster pump, intercoolers for the blowers, inverters for the electric instruments, electric dopers for engine-starting, integral control locks, and cigarette (cigar ?) ashtrays in the pilot-seat armrests.

For us – brought up on Blenheims, Hurricanes, Wellingtons and the like – the Liberator was a phenomenal affair. It was a flight engineer's dream of switches, warning lights and check lists. For every 30 minutes in the air we seemed to spend an equivalent time going through the cockpit drill, setting up the boost for each engine separately on the turbo-blower controls, creeping along the runway to get the nosewheel straight, and unlocking and checking the control surfaces.

The business of flying it was quite incidental and relatively unimportant, so long as the weather was reasonably good. When this went sour on a test or ferry flight, with no radio or radio-navigation aids, one felt trapped and in need of every ounce of grey matter. If the cloudbase came down you really had to think very hard indeed.

The Liberator was not an aeroplane which could be flung merrily round in a tight circuit of the first aerodrome to be seen. There was a lot to be done before it could be aimed at a runway on finals and–even if the runway was recognised and known to be long enough–you could lose sight of it altogether whilst making the necessarily wide circuit and working through the cockpit drill.

During earlier, do-it-yourself, days at our maintenance unit we had one or two fairly nasty rides when we were learning to cope with B-24s and to test them at the same time. There was, for instance, the day when the nosewheel shimmied violently just before take-off and could afterwards be locked neither up nor down. That day was saved by the flight sergeant rigger who was riding with us; he went down into the bowels of the aeroplane and fiddled with the nosewheel-door mechanism while we, at his shouted instructions, worked the undercarriage lever up and down. He fixed it (we never knew how) and came up beaming.

But our nastiest experience had nothing to do with technical difficulties–it was the result of a combination of inexperience and pride. We just didn't know what a handful a big and complicated aeroplane could be without navigation aids in marginal weather conditions. Our Hurricanes and Wellingtons and our knowledge of local topography had made us a little too cocksure.

It was one of those grey days on which the height of the cloud ceiling could not be assessed and, even though spring was round the corner, the weather was cold and there was a haze of snowy drizzle. The Liberator had been made ready for test, there was nothing else on the morning's work programme and we didn't want to disappoint the ground crews and engineer officers. Warmed by our elevenses the chief test pilot and I collected our crew and went out to "have a look".

By the time we had climbed aboard and settled into our seats the sleet-drizzle had turned to fine snow, but we went through the long-drawn-out process of preparing for flight and running up the four engines. Having got so far, it seemed that we might just as well taxi the Liberator out to the leeward end of the runway and see what the weather was doing there.

When we had got to the end of the runway it was snowing quite heavily–but it seemed a pity, after all the trouble that had been taken, not to go for just a little test flight. Besides, the man in the chequered box was impatiently giving us greens with his Aldis lamp and we would feel such idiots as we taxied all the way back to dispersal. The chief test pilot turned on to the runway and lined up for take-off.

As soon as we were airborne it was very obvious that we were not going to make much useful progress–except towards the Station mortuary slabs–by staying airborne for a moment longer than was absolutely necessary. No sooner had I completed the post-take-off cockpit drill than, at a gesture from the chief test pilot, I started to go through the approach drill. Meanwhile he edged the Liberator round in a low-level circuit using his knowledge of the landmarks, as these came into view through the gloom, to get the outfit lined up for an approach to a runway which couldn't be seen.

His judgment of the 360-degree turn was pretty good, but even at 200ft we were still letting down towards an invisible aerodrome. Then, dimly through the murk, we saw a flashing green light. The runway controller was doing his very intelligent best for us.

The light was way over to the right and we did what should never be done with this kind of aeroplane–a fairly tight S-turn towards the runway threshold. We were high and fast and as second pilot I also did what should never be done–I cut the throttles before we touched. The Liberator hit hard and the chief test pilot found time to give me a pained look as his back straightened while pressing on the brake pedals. I busied myself with the idle cut-offs for the outer engines, the cooling-gill switches and the flap lever. Nice to be home again. Two months later I was sent to Prestwick for this conversion course.

Near the runway threshold my co-pupil stopped the Liberator on the perimeter track while we changed seats for my last practice circuit in the first pilot's position. Laboriously we went together through the take-off drill.

Hydraulic booster and fuel booster pumps on; mixture controls in auto-rich; turbo-blowers set individually for each engine, as it was run up, to 47in. of boost, and revolutions adjusted to 2,700 on each of the four little spring-loaded switches; elevator one degree nose up, rudder three degrees right and ailerons neutral; gills closed; and flaps 20 degrees down.

I taxied on to the runway, straightened the nosewheel, unlocked the controls and checked them for freedom of movement. A green light shone from the runway controller's chequered box. Time to go.

The handful of throttles was eased forward with instinctive bias to keep the Liberator straight. We were on our way. "Twenty inches . . . twenty-five inches . . . thirty inches . . . thirty-five inches," my co-pilot sang out. My toes dropped down from the brake pedals as I felt the air loads on the rudders. "Forty inches . . . forty-five inches . . . I've got 'em. Airspeed . . . one hundred . . . one-oh-five."

I lifted the thing off and selected undercarriage up. My turn to talk now. "Blowers out . . . forty inches . . . twenty-five-fifty revolutions . . . flaps up . . . booster pumps off . . ." The Liberator settled into its initial climb–leaving us with the curious feeling of having done no more than a fairly competent engineering job of work.

Relaxed for a moment, I looked over the land and sea–lovely in Ayrshire's early summer.□

CHAPTER SIXTEEN

Fighting Fortress

*

FORTRESS

TEST PILOTS were not always popular people at RAF maintenance units. It was the job of these units and of the engineer officers, who (mostly) ran them, to produce equipped and serviceable aeroplanes in promised numbers. Testing was a necessary part of the routine, but the engineer officers could sometimes reach the well-controlled verge of tantrums when they suspected that a pilot was being over-fussy.

So it was that, for a week or two in the summer of 1942, I had been (like others before me) in the doghouse. Three times I'd abandoned the take-off with a B-17 Flying Fortress–and three times the technicians had found nothing to justify my complaint that the aeroplane "wasn't accelerating well enough to have a reasonable chance of being airborne before reaching the end of the runway".

I knew very well that the brakes were binding once the Fortress had reached a speed of about 60 knots, but it was no use saying this. The brakes were always free enough when the aircraft was jacked up for post-"flight" examination and the four Wright Cyclone engines always registered, on run-up (just as they did on take-off), the required inches of boost and revolutions per minute.

The senior engineer officer (or chief technical officer, as he was later known) may or may not have suspected me of malingering, but he certainly doubted my judgment. For that I could hardly blame him; I had tested only two B-17s since completing a conversion course on the type. No one else at the unit had flown the Fortress, so no second opinion was immediately available.

A few days after the third abandonment of take-off I saw an unexpected, but familiar, face at dinner in the mess–that of my instructor on the conversion course. Evidently the help of 41 Group headquarters had been enlisted and their B-17 specialist had been seconded temporarily to my unit.

Next morning the rogue Fortress was made ready for the fourth time. With appropriate apologies the specialist signed the authorisation book himself and we climbed aboard. I'd given him all the facts and, as I ran through the pre-take-off drill from the second pilot's seat, I could only hope that he wouldn't feel compelled to play a "Saint George for England" act and roll us into an untidy ball at the end of the short runway.

I should have had more confidence in his good sense. As before, the Fortress accelerated normally during the initial stages of the take-off, though the tail came up, I thought, a little too readily. Then it just stopped trying. As my left hand

was starting to itch for a handful of closed throttles, the instructor pulled them back himself.

This time there was no question of taxiing the Fortress back for another possibly ignominious ground check. We got it off the runway and on to the perimeter track and there it stuck, with solidly gummed-up brakes. The specialist returned to his unit and I got on with the rest of the day's test-flying programme.

Meanwhile the B-17, temporarily immovable, was causing an inter-unit quarrel of inprecedented fury. The pupils of the operational fighter training unit could not, with their Spitfires, safely negotiate the obstacle by leaving the perimeter track, though we, of the "lodger unit", could churn fairly happily through the rutted mud in our Hurricanes and Wellingtons. So the words burned the wires: "Get that damn Fortress (or whatever it is) off *our* perimeter track . . ." and so on. For once I was glad to be no more than a poor benighted (if relatively elderly) pilot officer in the RAF's general duties branch.

In fact, the discoveries made when the B-17's brakes were finally dismantled and examined were possibly worth more to the war effort than a few hours of OTU flying time – but that's another story.

□□□□□

From the pilot's point of view the Fortress was a great device – once those of us who had been brought up on do-it-all-yourself British military aeroplanes had got used to the idea of a team operation on the flight deck.

Like most American aeroplanes it seemed to have been designed for pilots with short legs and long arms. On early flights I had felt pretty well out of control

of the situation during take-off, with my left arm stretched at full length on the control wheel while getting the tail up; with my right arm stretched out at 30 degrees to take the throttles through to maximum boost; and with my feet treadling away for dear life on the brake-cum-rudder pedals.

The throttle action, too, was a bit abnormal, though we learned to like it. The four levers were formed into a kind of broken "three-barred-gate" in which the top bar operated the outers, the bottom bar operated the inners and the four-piece centre bar operated all of them. The trick was to learn how to give the throttles the necessary take-off differential while holding this centre bar in the only practical way, with the palm upwards.

But, once the pilot had got used to the necessary and somewhat cumbersome manual dexterities involved in its control, the Fortress behaved rather like an oversize, four-engined Anson of the flapless pre-war vintage. Of course the B-17 *had* flaps, but, at least when one was first learning to cope with it, they didn't seem to provide very much in the way of drag.

So one had to learn to hold really accurate speeds on the approach–which was something new for pilots accustomed to aeroplanes in which full flap could be depended upon to kill all excess speed as the runway threshold came up. During my first circuits with the Fortress I made three successive overshoots at Burtonwood before the instructor took pity on me and decided that the remainder of my dual could be completed at nearby Woodvale, where the runways were longer and where there were none of the embarrassments of industrial smoke and surrounding urban paraphernalia.

To British minds, eyes and hands, the B-17 had some oddly incongruous detail control features. The brakes and engine-cooling gills, for instance, were operated by hydraulic power, but the undercarriage and flaps were worked by electric actuators. So the most important of all ancillary controls, that for the undercarriage, was a tiny little switch under a safety cover on the starboard side of the vast central control console and almost out of reach of the first pilot.

The Fortresses which we did our best to test in 1942 and 1943 were (mostly) destined for a comparatively peaceful, if arduous, life with RAF Coastal Command on North Atlantic submarine-hunting patrols. I didn't learn about the real fighting Fortresses until 1944.

In the summer of that year a dozen of them had been diverted to our aerodrome after a daylight raid. At least two came in with feathered propellers, so I might have guessed what I was letting myself in for. But, keen to see the inside of a B-17 again, I motored around the perimeter track and climbed aboard the first in this beautifully parked row of aeroplanes.

As I looked at the shambles inside the Fortress and at the ankle-deep shell cases, I understood something of the final effort which must have been called upon by the crews in parking those Fortresses in such a carefully disciplined line.

These experiences with American aeroplanes were comparatively rare at RAF maintenance units. More than 90 per cent of the test flying was in more familiar British aircraft–such as the Miles Master in the narrative which follows.

Master IIIs in formation showing the retractable windscreen for the instructor's cockpit.

CHAPTER SEVENTEEN

Minor Emergency

*

MASTER

EVERYTHING about the behaviour of this Miles Master II trainer had been quite normal during take-off and the first part of its climb. It had put up the usual fey, high-spirited performance, which was typical of any Mark II or III, during the first half-dozen or so seconds while accelerating to what would now be known as "rotation speed".

The slipstream-sensitive elevator had raised the tail just a bit too high at first before forward pressure on the stick could be reduced; and it had tried to rush off to starboard in spite of full left bias and hard rudder pressure. The sing-song note of the Bristol Mercury engine had risen to the expected high-pitched level and had then settled down as the propeller's constant-speed unit had taken control.

In short, this seemed to be the beginning of just another routine flight–the fifth during a day in which the RAF maintenance unit's test programme had consisted exclusively of Master IIs and Master IIIs with Pratt & Whitney Wasp Junior engines.

After lift-off there was the usual slight difficulty in operating the undercarriage lever. This, sprouting out behind the control unit on the left, had to be pulled back and lifted over a safety stop before being moved to the "up" position. Then we were really on our way. The throttle was brought back to rated boost and the constant-speed lever adjusted to give climb revolutions.

At that moment, without any warning, the engine note dropped sharply through a couple of octaves. I moved the constant-speed lever forward again and glanced at the revolution counter. It showed no more than a feeble 1,300rpm or so.

The Master continued to clamber along at a steadily reducing airspeed as I tried to gain a little height while experimenting with various positions of the CS control. After moving it several times through the length of its quadrant, without any appreciable change in the engine note or the indicated rpm, it dawned on my slow-moving brain the that propeller control system had failed. The constant-speed unit was now, in fact, living up to its somewhat confusing name by providing one unalterable and more or less constant, but regrettably low, engine speed.

Luckily the Master, as a non-operational aeroplane, was fitted with a somewhat old-fashioned propeller installation in which the change of blade angle was restricted at each end of its range of movement. So the coarse-pitch setting, to which the failure of the cs unit had evidently committed the propeller, was at least not so miserably high-geared that further airborne progress was impossible.

Eventually, as the Master was levelled off from a laborious, near-to-tree-level climb, the revolution counter was registering a magnificent 1,450rpm or better. By then a wide half-circuit had been completed and we were reasonably well-placed for a turn into wind towards the runway in use–though this runway looked a very long way off and hardly below the dizzy altitude of 800ft which the Master had managed to attain.

Nobody at wartime RAF maintenance units or at factories used radio telephony. There was a shortage of frequencies, or crystals, or both. So there was no means of warning the airfield controller, in his chequered box, that I had a minor emergency on my hands. And, whether he had given me a "red" on his Aldis lamp, or fired one from his Very pistol, I would still have had to come on in now that the undercarriage had been selected down.

Perhaps it was as well that the controller was as powerlessly uninformed in his box as I was in the not very serviceable Master. In the event, he gave me a steady "green".

□□□□□

The Master was a pleasant enough aeroplane to fly, with brisk reactions even if the ailerons *were* a bit on the heavy side and the elevator and rudder a shade too light. Test pilots at RAF maintenance units–where there might be a dozen different types to fly week by week–were necessarily the more objective of its critics.

We did not, for instance, like all that plywood and glue and were duly comforted by the feel of our parachute harnesses and packs when, perhaps, the rudder pedals moved of their own volition or the ailerons twitched at high rates of roll. But we did not worry very seriously and our Masters, however ill-treated, always stayed in one piece.

Our thinking (if any) might have been conditioned by that apocryphal story of the factory pilot who complained, on successive production re-tests of one particular Master, about a faint but persistent rattling noise in the port wing. Eventually the flight-shed engineers were encouraged to open up the ply covering–glue, staples and all. They found–or so the story went–a plate, a knife and the miserable cadaver of what had once been a Spam roll.

My first experience of a Master–apart from a second-pilot ride in the back seat during a test flight out of Woodley a year or so before the war–involved the delivery in 1940 of a very battered and grubby Rolls-Royce Kestrel-engined Mark I from White Waltham to Martlesham Heath.

This was after I had learned to cope, in a ferry pilot's limited fashion, with Hurricanes, Spitfires, Henleys and Defiants. So the experience was not, as for the Master's normal pupil pilots, my first with a full-powered single-engined quasi-military aeroplane. It was just a job–and I was glad to get it done and to be ferried back, in an ancient Airspeed Courier, to the home base before nightfall.

Real knowledge of the Master, in its developed variants and as the Martinet target-tower, was to come much later. Then, as a still-not-very-experienced RAF

test pilot, it was possible to appreciate its cockpit arrangements, with the rational grouping of essential controls; the good cruising view over the nose; and the way in which the large flap area helped to make accurate approaches and landings so comparatively easy.

Like all aeroplanes, the Master had its little peculiarities. Those not very well matched controls, for instance, made accurate aerobatics, of a real kind, a little harder than they might have been. I was no aerobaticist, anyway. It has been all I could do to get by in pre-war years in a Tiger Moth during a half-completed instructor's course. But I did learn, during that and an earlier series of lessons, to recover promptly from spins after too-slow half-rolls-off-the-top.

So, when the occasional maintenance unit engineer-officer passenger in the Master asked for something more than a normal routine, I used to cheat with a manoeuvre which was very easy and in which nobody could have gone far wrong. This was, in fact, no more than a heavily barrelled roll–but it was most impressive to the uninitiated. You started to pull the nose up and then just applied hard aileron while holding the backward pressure on the stick and keeping the nose from yawing, one way or the other, by careful use of the rudder. The Master would whisk over once–or twice if you had plenty of speed–and would end the performance with the nose still comfortably above the horizon.

The whole performance was completed without any of that uncomfortable negative-gravity nonsense. I was not one who liked the Sutton harness to be biting into my shoulders while my feet floated away from the rudder pedals and the air of the cockpit was filled with dust, metal swarf, the contents of my pockets and any incongruous odds and ends which might have been left behind during the daily inspection. If the ailerons twitched the stick during the roll, one just eased the lateral pressure a bit and tried not to think too much about wood, glue and aerodynamics.

In fact we, as test pilots, had our private little worry about the Master's

ailerons. Riggers were in the (no doubt well-informed) habit of glueing little battens of wood *under* the trailing edges when lateral trimming was necessary. We knew, from previous experiences, that this could be dangerous because of the risk of excessive aileron upfloat–but accepted, not too happily, the fact that it was the accepted "book" method of re-trimming.

All the same, we looked on this advanced trainer very much as a pleasant and relatively harmless plaything–to be used, in preference to anything else, for single-passenger ferrying trips and for weather checks before the start of the day's work. Once or twice ("for experience") I let some keen technician take the place of honour and flew it from the back seat–marvelling afterwards at the skill, patience and rubber-necked dexterity of the instructors who rarely had the chance of sitting anywhere else in the Master.

□□□□□

Even during that fixed-coarse-pitch circuit I found it impossible to believe that, with a Master, the situation could be in any way really dangerous. All I had to do was to make sure that there was plenty of height and speed–but not too much of either–during the final few hundred feet of the approach.

All went well except for one unforeseen personal disaster after the landing. Coarse-pitch taxiing was clearly giving the Mercury engine a task which it could not accept–so I parked the Master on the compass base, to be collected later, and took a short cut on foot across the aerodrome. A hard winter had made a quagmire of the grassed areas. For some reason I was wearing my best blue and the commanding officer of the station had chosen that evening to re-institute a formal peacetime function in the mess. My mud-impregnated trousers were ignored with near-desperate politeness. □

CHAPTER EIGHTEEN

Prospective Catastrophe

*

HAMPDEN

THE CLOUDBASE WAS SOLID at about 2,000ft and the air was filled with a thin drizzle which, with the smoke carried on a light easterly wind from the industrial Midlands, reduced visibility to a couple of miles or less. A strictly limited amount of simple functional flight testing had been possible on that day in March 1943, and – with appropriate reservations in their reports – the pilots of our maintenance unit had so far flown everything that had been offered for test.

This MU was a "lodger" unit at one of those very dispersed, hutted, RAF camps which had been laid down in a hurry when the worst had finally been foreseen in 1938–39. The mess was a mile or so from the aerodrome, the primitive sleeping quarters were at least as far away again and a heavy, standard-issue bicycle provided the only (if healthy) means of transport between the three points. I had only just returned from a fortnight of attachment to another maintenance unit in Northern Ireland. There, the relatively easy test work, and the simple little ferrying trips to and from pleasantly named places like Murlogh and Ballywalter, had been combined with an off-duty life which included untroubled evenings in Belfast and the consumption of real shell eggs instead of those wartime decoctions of packaged powder. So I was feeling not a little depressed.

Now, while stooging around at 1,500ft in Handley Page Hampden AD851 I was also very conscious, while recording the various facts and figures, of the nearness of the Wrekin – that sharply ridged little mountain with a summit not so very far below my chosen flight level.

Just as I remembered, feeling more cheerful at the thought, that I'd been invited out to dinner and bridge that evening to a stately home within bicycle range, there was a sudden change in the cacophony of sound, a muted crash of metal and a succession of violent shudders through the Hampden's fuselage.

Looking round from the cramped single-seat cockpit for the cause of all this disturbance, I saw (without really believing what I saw) that the starboard powerplant was now no more than a basic and very naked and undernourished-looking Bristol Pegasus radial engine. The main cowlings had simply disappeared. All that was left in sight were the "works": exhaust ring, cylinder heads, valve gear, sparking plugs, intake elbows and cooling gills.

Instinctive reflex action had closed the throttles for me as soon as the rumpus had started and now, as the airspeed dropped, the buffetings began to die down. There was no one else aboard the Hampden with me and it was impossible, from the cockpit, to look back at the tail. I guessed that the airflow behind the uncowled engine might have got itself into something of a tangle, and that this could account for the more violent disturbances at the higher speeds. But the rudder pedals continued to twitch; this fact, I uncomfortably supposed, might be the result of some structural damage at the rear end. Well, we should see in due course. Keeping an eye on the starboard engine's cylinder-head temperature gauge, I started to increase power.

At that moment I glimpsed some rocks and tree tops disappearing under the port leading edge. Obviously they must be part of the structure of our old friend the Wrekin–which had been forgotten during all the excitement and (regrettably) the loss of height. At least I knew now where I was and turned on to a reciprocal heading for home.

When the airfield's runway system came into view through the murk on the port side I decided, in a sudden accession of common sense, that it might be a good idea to do some low-speed checks before coming in for the landing.

□□□□□

Luckily I knew the Hampden–and its ear-splitting Napier Dagger-engined blood-brother, the Belfast-produced Hereford–fairly well. I had first flown one out of its manufacturer's airfield at Radlett in November 1939, and had delivered it, miraculously intact, to the maintenance unit at Cosford, near Wolverhampton.

Unlike most bombers, the Hampden had a narrow fuselage, with no room for side-by-side seating, and the cockpit was strictly functional. An advantage, however, was that the pilot's field of view was almost unlimited from wing-tip to wing-tip.

Since then I had put in a few dozen hours in the two variants and had had one or two displeasing but educative experiences in the process.

Early in 1940, when the weather had closed in during a delivery flight, I had force-landed a Hampden on the little aerodrome just outside Worcester. A couple of days later I had flown it out again from a tail-in-hedge, brakes-to-full-power take-off. Later that year, bringing a Hereford over from Belfast to a maintenance unit in the Cotswolds, I had learned by accident something about its innocuous low-speed characteristics.

Although I didn't know it, the airspeed-indicator system had failed during the descent and the needle of the instrument had stuck at about 120mph. The Hereford had settled in quite gently as I started to hold off for the landing. It touched down outside the aerodrome boundary, bounced neatly over a slit trench and landed itself without fuss in the proper place.

By later standards the Hampden would seem to be a fairly strange affair – though it was natural and right enough in its era. Its large control half-wheel filled the single-seat cockpit and impeded efforts to reach the setting knob for the directional gyro. Hydraulic power for undercarriage and flaps was obtained by moving a so-called "power bolt"; this was on the pilot's right and looked much more like some cockpit-hood jettisoning device.

But it was a delightful aeroplane to fly. The aileron control was brisk and tight turns were a special pleasure because you could see from wing-tip to wing-tip. If you were prepared to take trouble with the business of landing, it offered considerable satisfaction to its pilot. Of course it could be allowed to wheel itself on, to complete an arrival like that of any good-tempered kangaroo, but an early check and a really firm final hold-off produced the kind of landing which made one's day.

Now, with all that uncowled machinery on the starboard side, I could hardly expect it to be quite so forgiving. The slow-speed checks completed, I settled cautiously for a minimum speed which might safely be maintained on final approach to a wheeled-on landing.

Looking round the Hampden after it had been taxied in, I could only wonder that the effects of this near-catastrophe had been so mild. The departing cowlings had cut slits in the skin of the centre-section and the starboard fin, and had neatly chopped the mass balance off the starboard rudder – which explained the twitching rudder pedals.

Two days later a farmer brought in the battered remnants. One look at them was enough. The cowlings had been fitted with their fasteners held on by the single rivets used by the manufacturers to keep the parts together during delivery. These fasteners nearly always had to be re-positioned and re-riveted, but this time the cowlings had gone into place without adjustment and no one had thought of putting in the extra rivets.

It was just as well, perhaps, that weather conditions had restricted the test work on that day to a few simple, functional routines at normal speeds.

CHAPTER NINETEEN

Tricycles need Tarmac

*

BOSTON

WHEN THE Douglas Boston became firmly airborne it should have been obvious to anyone but a moron that there was something dangerously queer about the operation of its hydraulic system. Undercarriage "up" had been selected immediately after take-off, but minutes had gone by without any sign of retraction and the operation had finally been completed by hand pump.

Only when I had run through most of the RAF maintenance unit's flight-test routines, and the time had come to check flaps and undercarriage before going back to the aerodrome, did it begin to dawn on me that the delayed hydraulic action and the need for hand-pump support might have some serious functional significance.

Sure enough, there was the same interminable delay after the undercarriage had been selected "down", and this time, when the hand pump was used to hurry the operation along, it seemed to have very little on which to work. It moved all too freely. I wondered what my technician passenger–incommunicado as he was in the aft section of the Boston's fuselage–was thinking about our successive and apparently pointless circuits and whether he could see what was happening to the main undercarriage legs.

Eventually the hand pump–operating, it seemed, on no more than a mixture of hydraulic-fluid froth and thin air–produced the long-awaited undercarriage-down indications. That, at least, was something–an essential start to the business of getting the Boston safely back on the ground.

I would have liked some flap, too; but that, apparently, was too much to expect from an exhausted hydraulic reservoir. It dawned upon me, too, that if there was no hydraulic pressure available for the flaps, then there would be none for the wheelbrakes either.

Well, the situation would at least provide a splendid opportunity for the demonstration of my theory that, even with nosewheel aeroplanes, aerodynamic drag was more effective than braking action when a really short landing had to be made. The tricycle undercarriage was, after all, an unusual thing in England even as late as 1943, and private demonstrations of this important kind were not normally possible at Government expense–win or lose.

I motored the Boston in on a straight approach and as slowly as I dared; cut the throttles just before the runway threshold came up; held the control column against my chest as the main wheels touched; and kept on holding it there.

ABOVE: *RAF Boston III light bomber*. BELOW: *Instrument panels and controls of an A-20 Havoc, the USAAF version of the DB-7.*

The essential part of the landing run was certainly short and the speed had dropped to no more than a fast taxiing pace before the Boston had covered half the available length of the runway. But it continued to roll gently along under the power provided by the two Wright Double Cyclones as they "blip-blopped" noisily away at the idle throttle setting.

How could the thing actually be stopped? Obviously, there was only one practical way. With a burst of port throttle I bowled it over the runway shoulder and on to the rain-sodden turf and moved the mixture controls to cut-off.

The rightness of my patent theory had not, I regretted, been finally proved–but at least Boston W8259 was still in one piece and other people would now be able to use the only into-wind runway.

□□□□□

The P-70, or DB-7, or Boston, or A-20, or Havoc–however the different variants might be remembered–was very much a pilot's aeroplane. It had, like all American aircraft, a few quaint cockpit ingenuities; but it was essentially a flying machine rather than merely the product of some effective technical achievement.

I had been told of its good qualities and had digested as much as I could of the handling notes, but I had been frightened enough when, nearly three years before, in July 1940, I'd scrambled out of the Air Transport Auxiliary taxi Anson at Speke Airport, Liverpool. There I was expected to collect the first DB-7 I'd ever seen and to deliver it safely to the maintenance unit at St Athan in South Wales.

My experience with tricycle undercarriages had been limited to a few short flights in pre-war light aeroplanes like the little Dutch Scheldemusch, the American Stearman-Hammond and the experimental Hornet Moth–which I broke, without knowing it, on the first of several successive landings. The DB-7, moreover, was a grown-up aeroplane and it had technical features which were likely to be a good deal more difficult to master than its nosewheel undercarriage.

But at least I did not imagine, after the sad experience with the Hornet Moth, that tricycle undercarriages had been invented so that you could fly the aeroplane straight into the ground. I'd learned that a modicum of hold-off was advisable and even necessary.

For the rest, this strange aeroplane was obviously going to demand very special attention. Not only were the controls and instruments unfamiliar, but this early DB-7 (AE463–one of the batch which had been diverted to England after the fall of France) had all its instruments calibrated in metric units and its controls and instruments labelled in the French language. The fuel gauges, for instance, were marked in litres and the engine manifold pressures were unintelligibly calibrated in *pièzes*. I was relieved to find that the engine-control linkage had been modified for the benefit of normal (Anglo-Saxon) pilots. The throttles opened forwards and not backwards as was the custom in French-built aeroplanes of the inter-war era.

When–after a long and (for the ground crew) boring period of cockpit inspection and slow-motion starting of the two Double Cyclones–the chocks had been waved away, I was already learning something about the advantages of the tricycle undercarriage. As soon as the DB-7 was on the move it became obvious that the more arduous parts of the business of flying had been changed for ever.

Not only could you see where you were going, but you were in complete and reasonably accurate control of the situation. Within a few seconds I'd learned the trick of holding constant power and *driving* the aeroplane towards the take-off point, using the then-unfamiliar foot brakes to control the speed and direction. I had yet to learn about the advantages of keeping the (then newly invented) control lock engaged, so that the column and rudder pedals were not flopping about in the usual irritating way; that trick was to be appreciated later.

The hour's flight to St. Athan might have been a reasonably relaxed one if I hadn't tried to play tunes too conscientiously with the rather complicated multi-tank fuel system. Few of the 60 minutes of the flight were spent usefully in finding out about the DB-7's characteristics; they were devoted instead to the unsatisfactory business of converting litres into gallons by mental arithmetic and in trying, unsuccessfully, to make the various port and starboard fuel tank empty themselves in some rational way.

In fact, I was still turning fuel cocks this way and that, and watching those litre-reading gauges, when Cardiff and Barry and the aerodromes at Llandow and St. Athan were already coming up over the nose.

My pre-conceived ideas about tricycle-landing techniques may have been dominant in my mind–or perhaps I was still worried about those litres and *pièzes*. Anyway, I put the DB-7 down a little on the fast side at St. Athan and had to brake rather hard. The station commander–who had come out of his office to take a look at this extraordinary concept of an aeroplane–glowered when he saw the scars which had been cut through his beautiful turf.

Maybe he knew then that tricycles need tarmac and that the days of grass aerodromes were numbered. □

CHAPTER TWENTY

Entomological Interlude

*

MOSQUITO

WITH UNDERCARRIAGE DOWN, flaps at approach setting and a comfortable whiff of supporting power from the two Rolls-Royce Merlin engines, the Mosquito II was sliding smoothly along a gently turning descent path. Everything felt and looked right and I was enjoying the rare feeling of satisfaction which accompanies the conviction that (for once) there would be no need for any changes in power and that the rate of turn was nicely designed to line up with the runway.

I glanced again at the airspeed indicator–and then looked at it more carefully in astonished disbelief. The needle, which had previously been steady, was now making slow, erratic sweeps between the 80 and 200mph indications on the dial.

Something was obviously very wrong. I opened the throttles to plus nine, selected undercarriage up and climbed away. Safely above the circuit height of the Airspeed Oxfords from the advanced training unit using the same aerodrome, I took another good look at the ASI. The needle had now stopped making its meaningless sweeps and was settling slowly back to the zero equivalent on the scale.

Oh, well, other people had landed Mosquitos without the help of an airspeed indicator and had probably done it in very much less favourable conditions. Unfortunately, though, our runway was only 2,700 ft long and there wouldn't be enough room to play the situation out with the usual high-speed approach and screaming wheeler. I thought for a moment of going to another RAF station farther north where there was a 5,000ft runway, and then decided that this would cause an unnecessary disruption of the work at our maintenance unit.

In fact, everything turned out reasonably well–though the man in the chequered box must have wondered why this supposedly experienced test pilot was making such a complicated business of his normal job of work. The first approach was a bit on the fast side and it was obvious, when starting to hold off, that we were not going to fit into that 2,700ft of tarmac, but the second could hardly have been better. The Mosquito settled firmly on three points with a familiar crackle from the Merlins' exhausts as the power was finally cut about 200ft inside the runway threshold.

The cause of the failure was ludicriously simple, though of more interest to entomologists than to engineers. Some insect had chosen, during that summer of 1943, to lay its eggs in this particular Mosquito's static-line system. In due course the eggs had hatched out and hundreds of (no doubt mystified) premetamorphic grubs had crawled through all the pipelines, making nonsense of the pressure differentials on which the workings of the ASI depended.

After that the engineers were supplied with flagged plugs to put into the static vents of Mosquitos while they were on the ground. I wish I could think that the difficulties and "incidents" reported by pilots could always be guaranteed to produce corrective measures with a similar order of promptitude.

At that time I'd put in, maybe, eight hours of test flying on this effective and delightful, but sometimes mildly temperamental, device. During the next ten months or so I was to learn a great deal more about the different versions of the Mosquito.

At maintenance units, where tests of one type were interspersed with those of half-a-dozen others, pilots learned to simplify their attitude to type variants. For us there were just two kinds of Mosquito: the slightly claustrophobic bomber/PRU affairs, with their shallow vee-screens and control half-spectacles (which tended to inhibit any very dashing manoeuvres), and the fighters and fighter/bombers, with their deep, flat bullet-proof screens (providing a much more reasonable view of the world at large) and with their control sticks (which were far from inhibiting).

Luckily we had never heard, or not seriously listened to, the stories about the Mosquito's "vicious take-off swing". We were not, consequently, infected by the near-panic which, leading to over-cautious take-off handling, was sometimes the cause of so-called ground loops. Of course the Mosquito–like any other high-powered multi-engined aeroplane–*would*, if permitted to do so, swing quickly and

VL618

finally to its own destruction. But, if the two Merlins were opened up quietly, leading with the port throttle–and if you waited patiently until the air loads could be felt on the rudder pedals before taking the throttles through to maximum boost–the Mosquito's take-off behaviour was as untroubling as it was exhilarating.

Landings, though, could sometimes be a bother when conditions were unfavourable. The only serious ground loop during my period of duty with a Mosquito-holding maintenance unit was made on landing, but it was certainly academic in its perfection. While a new test pilot was making his first solo on the type, a strong wind had quite suddenly shifted its direction. He made a smooth touch-down on the one–now very out-of-wind–runway and the Mosquito had run straight for a few breathless seconds with a plume of deepening smoke coming from a heavily braked port tyre. Then it had made a twinkling pirouette into wind and had ended up sliding backwards over the wet grass. That particular Mosquito needed a new pair of undercarriage legs.

Except when flying alone and at reasonable altitudes we tried not to think too much about real emergencies and the possibility, or otherwise, of successful abandonment. The Mosquito's monocoque fuselage–made in two lengthwise halves, rather like those of the pod of a pea, and mated after most of the equipment had been installed–was not appreciative of large "cut-outs". The fighter versions had a small door on the starboard side (through which one might presumably escape successfully by diving out under the wing), but in the bomber/PRU variants the "door" was a floor hatch under the passenger–who would

DZ313

obviously need to be fairly well drilled in his departure actions if the driver was to have any kind of reasonable chance of survival.

In fact, we much preferred not to carry "passengers" on test flights – but all kinds of people were keen (goodness knows why) to fly with us and the second seat was rarely empty. In any case, many of our test programmes included professional checks of AI (air interception) radar and somebody had to come along to do this job.

On these radar test flights, with Mosquito XIIIs, we would try to arrange to go off in pairs, closing in on one another from different angles while the radar specialist sang out the ranges and relative positions ("two thousand yards slightly above, ten degrees to port") and we, in our turn, tried to judge their correctness by visual means. When no second aircraft was available for these tests, we would set off alone and search the sky for suitable "random" targets.

On one of these target-hunting flights I had closed steadily in on an American B-17 Fortress bomber (probably on its way from Prestwick after an Atlantic delivery) while the radar operator continued to report "nil" returns. Finally, when the Fortress seemed to be filling the windscreen, and I had began to get nervous about the possibility that someone on board it might choose to fire a playful burst, I told the radar operator to have a look for himself.

He looked. All he said was "Lordy", and gazed at me for a moment with a shocked but patient expression. Then he looked out again and, as the *cliché* goes, his "eyes lit up". Afterwards he admitted that he'd never seen anything quite so strangely beautiful as that scintillating mass of machinery poised, apparently stationary, against the turquoise sky.

But we had our difficult moments, even with such a pleasant and comparatively easy-to-handle aeroplane as the Mosquito. Sometimes (because the hydraulic system wasn't really up to it) the undercarriages couldn't be made to lock up. Occasionally we would find, after tab adjustment, that the ailerons were overbalancing seriously at lower speeds, or that they were "up-floating" dangerously during power dives. We were never really quite happy, either, about all that plywood and glue – though it certainly worked and the real Mosquito pilots (those who were being shot at) reported that the structure usually held together in a quite remarkable way after it had been hit. We complained, too, about the cordite fumes which sometimes filled the cockpit during air-firing tests, and about the contortions needed to reach the fuel cocks.

But who were we to dare to complain? Our test flights were made in daylight and we had moderately comfortable beds waiting for us after each day's stint of work.

□□□□□

Alongside the aeroplanes which we got to know so well, such as the Mosquito, the Beaufighter and the Master, this maintenance unit – No 27 at RAF Shawbury, Shropshire – prepared some strange oddments from time to time. There were, for instance, the Brewster Bermuda torpedo dive bombers which lay about the dispersal fields waiting for modification kits which never came, and the rare Armstrong Whitworth Albemarle. □

CHAPTER TWENTY-ONE

No Spice in Variety

*

BERMUDA · CLEVELAND · NOMAD

WHILE WAITING for a flashing green signal from the controller's Aldis lamp, which would allow the cumbersome dive bomber on to the runway for take-off, I suddenly discovered that there was a nearly impossible decision to be made. After completing the take-off drill at the end of the perimeter track I had locked the Sutton harness and then found that the undercarriage retraction lever could not be reached with my shoulders held back by the lock–as they should be for take-offs and landings.

So I had a ridiculous choice of alternatives. One was to take off with the harness locked back and then be unable to raise the undercarriage if, by any chance, the engine chose to stop. If it did so, the wretched thing might well go over on to its back as it hit the boundary fence and I stood a good chance of a broken neck. The other choice was to leave the harness free so that the undercarriage could be raised quickly in case of trouble–but at the risk of reducing my face to a sort of scrambled egg against the instrument panel as the outfit slid to a stop in the next field.

Luckily the controller gave me a flashing green, and immediately afterwards a steady green, before there was time to think seriously about this dilemma. I elected to take the scrambled-egg risk–in exchange, also, for a briskly efficient-looking take-off on the first test flight with one of the Brewster Burmudas (FF562) held by our maintenance unit.

Why, it might well be asked, should there have been this ten-second heart-searching about the possibility and effects of engine failure? Powerplants rarely failed even in the long-ago days of 1943. These Bermudas, however, were rather special creatures. A dozen or so were dispersed in the fields around RAF Shawbury in 1943–44. These awaited, among other modifications, one to the wing spar. In consequence there were strongly worded speed-restriction placards screwed to all the instrument panels.

More than 700 of these aircraft–originally US Navy shipborne SB2A Buccaneer dive bombers–had at one time, according to the records, been allocated to the RAF under lease-lend terms in a land-based, non-folding-wing version. In the event, they turned out to be unsuitable (to use a euphemism) for operational purposes and their manufacture had been abandoned after deliveries of a comparatively small number to the USAF (as the A-34) and to the RAF (as the Type 340 or Bermuda). Some of the British versions were later modified for

ABOVE: *Cockpit of the SB2A–4, the original US Navy version of the RAF's Brewster Type* 340 *Bermuda.* FACING PAGE, TOP: *Curtiss SBC-4 dive-bomber, known to the RAF as the Cleveland; Brewster Bermuda in RAF markings; Northrop A–17A light bomber, supplied to the RAF as the Nomad.* FACING PAGE, RIGHT: *Another view of the RAF Cleveland, head on.*

2669M

target-towing duties, but most of them were (in that delightful Service phrasing) "reduced to produce" in 1945.

In the meantime, we, the MU test pilots, had to fly them periodically, just to demonstrate the fact that they were still more or less in working order. None of us had any particular reason to question the reliability of the Wright Double-Row Cyclone GR-2600s with which they were powered–though, because of some peculiarity in the layout of the exhaust system, the engine made alarmingly about-to-fail noises at take-off power. Our vaguely felt discomfort was the result of the fact that the Bermudas, before we flew them, had been sitting about for weeks in the dispersal fields with their engines "inhibited" with preservation oils of which the Queen of Sheba had certainly never heard. So, perhaps, we can be forgiven for not being too brightly enthusiastic about the chances of take-off success from Shawbury's then-all-too-short runway.

We had another problem, too, with this monstrous single-engined bomber. At about the time when we first started to re-test them in rotation, our lords and masters at 41 Group headquarters had decreed that each taxiing accident would in future be the subject of a full inquiry and a possible grounding (or even a court martial) for the pilot involved. There had been too many such accidents at MUs and Group had decided that a tough line was necessary.

The Bermuda, with its big cowled radial engine, offered no forward visibility at all for its pilot. His only means of obtaining a view of the perimeter track or runway ahead was to "weave"–in other words, to taxi in a continuously snaking path, looking forward, at each "weave", over the port and starboard gunwales of the wide cockpit. Unfortunately the Bermuda also had a very wide undercarriage and rather harsh pedal-operated hydraulic brakes. So, when the Group dictum was announced, the test pilots insisted that there should be a man at each wing-tip during all taxiing movements. This was obviously a gross waste of valuable manpower and, so far as I can remember, the decree was rescinded in the case of the Bermuda–perhaps because the safety or otherwise of these aircraft was not considered to have any particular significance for the war effort.

Maintenance unit test pilots, like Air Transport Auxiliary ferry pilots, found themselves flying some fairly queer devices from time to time–and sometimes 'off the cuff", with no handling notes, or even experienced, first-hand advice for guidance. We had to find out about them as we went along. There was, for instance, the Curtiss SBC-4 dive bomber, or Cleveland as it was known to the RAF–though why it had a special type name, when there were never more than half-a-dozen in the country, is a question which, maybe, someone else can answer. In 1940 I ferried one of these extraordinary relics of an earlier era in US military aircraft development from Burtonwood, in Lancashire, where it had been assembled and given the number AS468, to the MU at Little Rissington in the Cotswolds.

The Cleveland was a single-bay biplane which, though it had a big-diameter Cyclone radial in the nose and a variety of centre-section struttage and other excrescences to provide a built-in headwind, also had a retractable undercarriage and fully enclosed, sliding-roofed accommodation for the crew. The retraction antics of the main undercarriage, with its complicated geometry, are about all I can remember of a trip which must, with such a strange aircraft, have been at

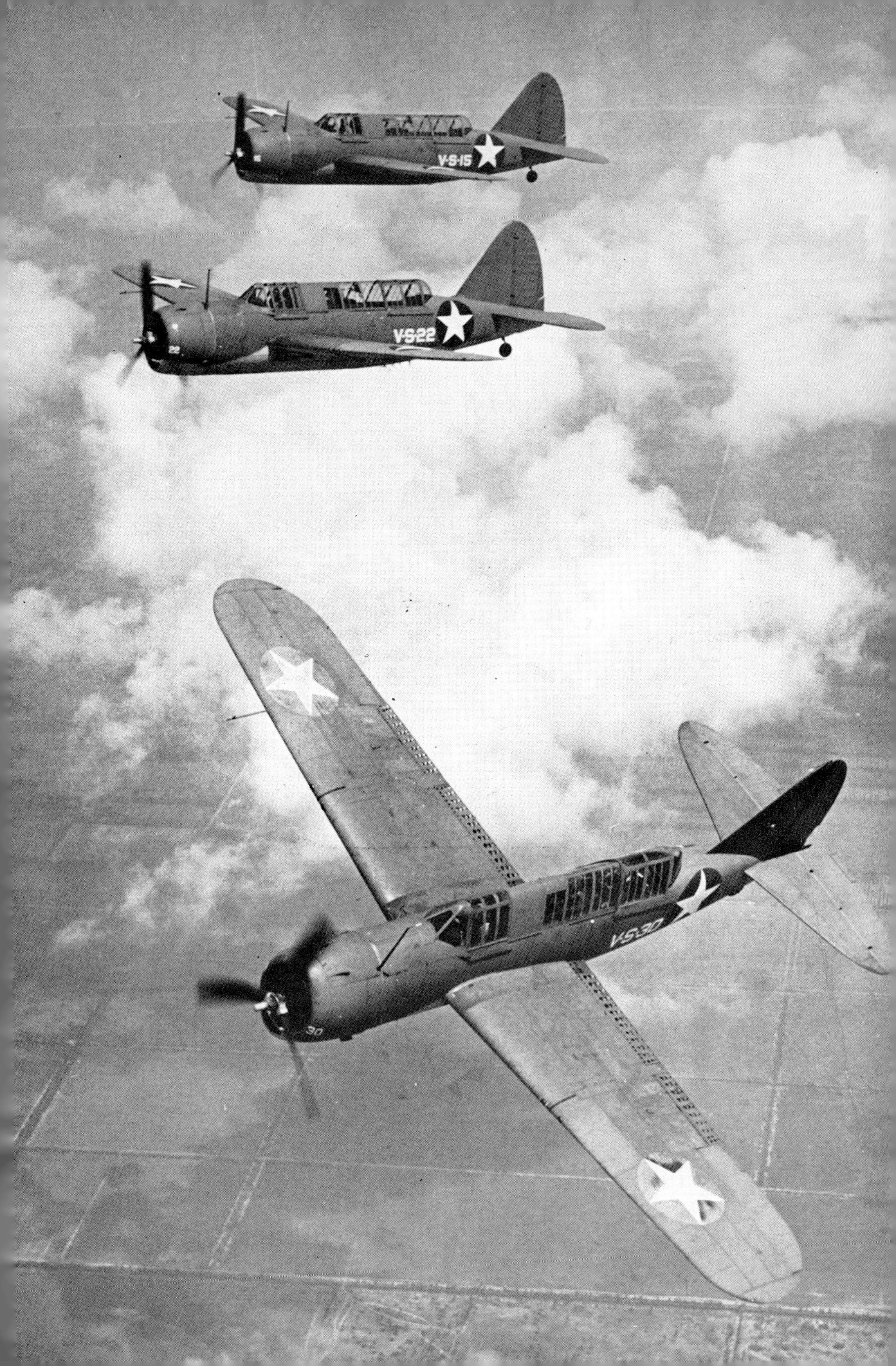
V-S-15
V-S-22
22
V-S-30
30

least mildly worrying. I suppose the fact that the Cleveland was a biplane, and therefore familiar to one brought up on biplanes, provided an irrational but comforting feeling of relaxation.

At Number 48 MU, Hawarden, Chester, in 1941 there were a number of Douglas (Northrop Division) A-17A Twin Wasp Junior-engined light bombers which we tested at intervals – and which we later delivered ourselves. To us they were just "our Northrops," easy-to-fly, single-engined, tandem-seater, low-wing monoplanes with retractable undercarriages, which had been, for no particular reason, parked at our unit and were duly maintained. The record says that they were type-named Nomad by the RAF; that they were assembled (like the Cleveland and many other US aircraft) at Burtonwood; that about 60 of them had been handed over to the RAF by the US Army Air Corps; and that most of them eventually went to the South African Air Force.

During their stay at Hawarden they provided us with a welcome opportunity for some reasonably relaxed test outings and also, in those earlier war days, some added and essential experience of the characteristics and curiosities of American aircraft. We were, for instance, accustomed to blaring horns and red-light warnings – but who would have thought of the (no doubt psychiatrically sensible) idea of using an electric actuator to make the rudder-cum-brake pedals tremble uncomfortably when the throttle was pulled back with the undercarriage still retracted?

That strange device in the Northrop was only one of a score or more of ingenious new ideas and systems – many of them now accepted as standard features – to which the test and ferry pilots had to become accustomed in 1940 and 1941 as US aircraft, some good and some bad, started to arrive in greater numbers and diversity. □

CHAPTER TWENTY-TWO

Preternatural Prototype

*

ALBEMARLE

As the thousand-foot needle of the altimeter passed through the figure "5" for the second time, I pulled the oxygen mask from my face and called to the flight engineer, asking him to check that the oxygen supply had been turned on and that my own line was connected. A few moments later his head appeared from down below and he gave a thumbs-up sign.

In 1943 the test pilots at RAF maintenance units rarely had need to use oxygen and found it difficult enough to obtain bottles and to get the system laid on for special tests. But there had been no problems of supply and installation in this case. One of several Armstrong Whitworth Albemarles had to be taken up to its service ceiling of 18,000ft in preparation (we were told unofficially) for a delivery flight, by a devious route, to Russia. Apart from later operational requirements in its use as a freighter, the delivery flight would involve some long-period cruising over high ground (the Caucasus?). There had been complaints of continued low engine-oil pressures and excessive temperatures during and after climbing to this phenomenal altitude, and this test was one of a series designed to check that oil-cooler and other modifications to the Bristol Hercules engine installations had put things right.

Not everyone knew then that the effects of oxygen-lack are insidious and are not appreciated as being dangerous (even if they are noticed) by those who have had no experience of them. The initial effect is somewhat similar to that of a fairly large dose of alcohol on a natural extrovert; everything one does is just splendid and there are absolutely no problems.

A few months before this test flight, when serving with another MU on a station which was occupied by a Fighter Command operational training unit, I had been given the chance of experiencing the effects of anoxia in a decompression chamber which travelled around the country on demonstrations. Breathing oxygen in the chamber, I had watched the capabilities of a fighter-pilot pupil, who was not breathing oxygen, gradually but surely disintegrate. His task was to write some simple words and to sign his name repeatedly on a pad provided. His writing slowly lost cohesion, until the note and the signature were no more than formless scrawls. He had been told to concentrate on this simple chore. Afterwards, when he discovered that–under instruction from the medicos in the fresh air outside the chamber–I had removed his shoes, he was effectively shocked into an appreciation of the effects.

And so, in my turn, was I. Even though I had seen, only a few minutes before, these results of lack of oxygen, I still thought that the symptoms could be recognised and that I would be able to adjust my behaviour accordingly, so as to be at least a little less obviously incompetent and idiotic. Throughout the test, I believed that my signature remained legible and that I would certainly notice if anyone started to remove my shoes. Nobody had, in fact, removed my shoes – but, as I discovered when back on oxygen, somebody had, instead, removed my wristwatch and the contents of two of my pockets. What I had continued to believe was legible writing turned out to be no more than a jumble of superimposed doodles.

There had been other tests waiting to be done after landing the Albemarle and I never remembered to ask whether the flight engineer's thumbs-up sign meant that my oxygen had now been turned on or that it had been on all the time. Anyway, I had felt (or imagined I had felt) a lot better and took the Albemarle up to about 20,000ft for good measure. Whether my suspicion had been right or not, the earlier experience in the decompression chamber turned out later to be very valuable when, at 21,000ft in a Spitfire, I started to laugh at the faces of the instruments and at the figures on my test pad. In the temporarily enjoyable drunkenness of anoxia a separate "me" remembered that this is what it had been like in the decompression chamber. I closed the throttle and put the nose down before searching down below in the cockpit to find (laughably enough) that the connection of the oxygen tube had come apart. With immense effort and concentration, giggling at the absurdity of the need to take such action, I reconnected the line. Normality, when it returned a few moments later, seemed rather uninteresting.

The Albemarle was, for a British aeroplane, unique in one important particular – it had a nosewheel undercarriage, or a "tricycle", as we called it then. It was also, in other minor ways, nearly as strange as its heredity. Originating in a 1938 bomber-reconnaissance specification, its design and manufacture had been transferred from Bristol (as the Type 155, with two Taurus engines) to Armstrong Whitworth (as the Type 41 with two Hercules engines). The idea by then was that it should have a composite structure of steel and wood, so as to conserve strategic materials (e.g. duralumin) and allow the production to be widely sub-contracted through various industries and its assembly to be completed by a non-aviation company. About 600 of them were built, mostly (after the first batch) as paratroop carriers and glider tugs – and in these forms they eventually went to war.

We could have had no knowledge of such later uses for the Albemarle when a few of them were being prepared in 1943 for delivery to Russia. It was just an aeroplane which had to be tested for general airworthiness and, later, for its fitness to make this delivery flight.

For me, it was a specially interesting device because of its nosewheel layout. In retrospect, it seems strange that such a relatively unimportant project should have been designed with this new-fangled device. But the fact was that the development of the contemporary twin- and four-engined heavy bombers had already gone too far for change. The time-scale of their design and production did not, in war-emergency conditions, permit inessential major modifications of this kind. Only with the application of jet engines and the chance for starting designs from scratch did the nosewheel become a necessary and logical requirement.

Nearly all my experience of nosewheel-handling techniques had been with

The Albemarle had the familiar British thumb-operated brake lever and differential steering to go with its castoring nosewheel.

American aircraft–the Douglas Boston, the Bell Airacobra and the B-24 Liberator. All these had pedal brakes to go with the castoring nosewheels. The Albemarle had the familiar British thumb-operated brake lever on the control wheel and differential rudder-pedal steering. Although I had eventually become accustomed to the American method of braking and ground control, the layout of the Albemarle was a pleasant change after the earlier struggles.

Even so, taxiing could be troublesome on narrow perimeter tracks with parked aeroplanes all around. As I remember it, the nosewheel of the Albemarle was strongly self-centring and a heavy jab of the brake lever, with full rudder, was necessary to get the thing turning. The wing tips could not be seen from the pilot's seat, so a crew member had to stand with his head in the elongated astrodome behind the cockpit. From this position the wing tips could be seen but we never worked out a satisfactory means of communicating the facts of the situation when negotiating narrow passages between, say, parked aircraft and the corners of buildings.

There was another minor but prospectively dangerous characteristic of the Albemarle. For some aerodynamic reason, it developed a slight nose-heaviness as it was lifted off the ground–presumably the effect produced as the mid-mounted wing left the ground cushion. The undercarriage lever was low down in the front of the throttle pedestal, so, when selecting undercarriage "up", the pilot was, for a moment or two, out of sight of land. Later we learned to trim the Albemarle a bit more tail heavy for take-off than recommended in the book, or to get another crew member, if such was available, to raise the undercarriage. But when I looked up again after selecting undercarriage "up" on my first take-off the Albemarle had lost several feet of altitude and seemed almost to be brushing through the undergrowth beyond the end of the short runway. Interestingly enough, the flaps were designed to be used as dive brakes in their full-down position.

I have often wondered what happened to the Albemarles which went to Russia and to what use they were put when they got there. A later story was that the Russians had accepted the aircraft as the only means of obtaining Hercules engines–which they badly wanted. Of course, our Albemarles did not actually carry labels saying "to Russia with love"–but our information came by usually reliable bush telegraph.

Although the Bristol Beaufighter is the central figure in the prologue to this book, this sturdy all-purpose engine of war deserves special, separate attention. With the possible exceptions of the Spitfire and Hurricane, the Beaufighter was my favourite aircraft because of its straightforward handling characteristics, ample power and impression of indestructability.

CHAPTER TWENTY-THREE

Amenable Monster

*

BEAUFIGHTER

THE ROUTINE post-modification test flight was almost completed. The only snags had been a slight left-wing-lowness at the neutral lateral trim setting and a tendency for the artificial horizon to "topple" too easily in turns. Each Hercules engine of the Beaufighter VI had been throttled back in turn and its propeller feathered. Now, apart from the checks which would be completed during the descent and final approach, there remained only the tests to make sure that the different possible combinations of fuel supply from the inner and outer tanks to the two engines could be selected. The take-off had been made, as the book required, on the inners, and we were now flying on the outers. Near Iron Bridge, on the Severn, the "Beau" was headed back towards the Wrekin, that familiar landmark in the area, and to RAF Shawbury. Then I twisted round in my seat to turn the fuel-cock handwheels on the left.

It was a beautiful, cool day in the late spring of 1943, with visibility of around 50 miles; so my mind was not, perhaps, as concentrated on the job as it should have been. The first inkling of trouble was communicated by my right leg. It ached and I realised that it was instinctively pressing very hard on the rudder pedal. When I had relieved the load on the leg with a few turns of the rudder trimmer and had had a good look round, it dawned on me that the port engine must have lost power or even failed. Thanks to the constant-speed propeller and continuing high, though now reducing airspeed, the engine revolutions, boost and oil pressure readings for the port powerplant remained more or less normal, but the indicated temperatures for the oil and cylinder heads were dropping steadily.

Only as I started to go through the engine shutdown and propeller-feathering drill did I discover what had happened. During the fuel-feed changes I had somehow managed to turn off the supply to the port engine, which was now just being windmilled round.

The test flight was completed with rather more than usual care and it was a sadder and wiser man who taxied the Beaufighter to the MU site from which it had been taken half-an-hour or so earlier.

Though mildly intimidating on first acquaintance because of its almost brutally powerful appearance and apparently complicated cockpit, the Beaufighter was, at least in its Hercules-engined versions, a satisfying and even delightful device when you got to know it well. The controls may have been on

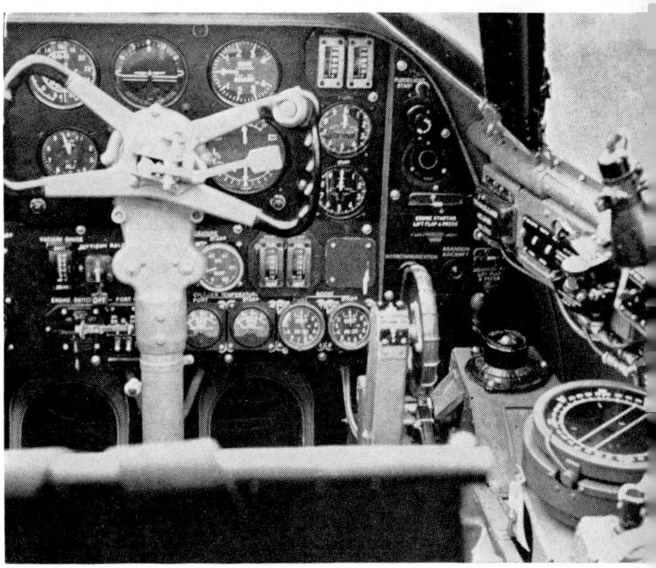

LEFT: *Beaufighter Mk I with original straight tailplane; Beaufighter Mk X with torpedo; Beaufighter Mk X target towing aircraft.*
ABOVE: *"The single-seat cockpit was a fairly untidy place. . . .".*

INCENDIARY
STOWAGE

the heavy side; but it handled beautifully, gave its pilot a near-perfect view ahead through the flat, bullet-proof screen and was surprisingly easy to fly so long as reasonable care was taken during the first part of the take-off run.

In its day, a deal of near-nonsense was spoken and written about the Beaufighter's "take-off swing". As I said when writing about the Mosquito, any aeroplane, especially one with the centre of gravity well behind the main undercarriage, will deviate from the straight and narrow if it is allowed to do so. An exaggeratedly stocky and powerful twin like the Beaufighter was no exception. Until adequate rudder control could be felt and seen, you had to keep it straight during take-off by using differential power–inching the throttles open individually while aiming at some distant object. Once there was full rudder control the throttles could be taken straight through the rated gate and the Beau accelerated along the runway like something on rails–and a very exhilarating experience it was.

Its reputation for directional control peculiarities was probably partly the result of the fact that the undercarriage legs would tend to come up in an uncertain and somewhat hesitant order, causing some pronounced and not very easily controlled yawings from side to side. The Merlin-engined Beaufighter II, in particular, felt sometimes as if it was going sideways for a few seconds during the initial climb. I remember reading somewhere, long after the war, that the long dorsal fin of later Beaufighters "cured the swing". Of course it didn't; the extra fin area just made it more controllable during the less easy part of the take-off run and the climb out. Anyone who wants to learn about the control of a take-off swing should discuss the subject with an ex-pilot of Short Stirlings.

Nevertheless, when things went wrong, the Beaufighter's aft centre of gravity could be a nuisance as well as, in the event, something of a blessing in disguise. On one test flight I found that the brake-operating air pressure had vanished mysteriously. After cruising around for a while in the hope that the needle of the gauge would move at least a little way from the zero against which it had settled, I realised that there was nothing for it but to try to make the most accurate, driftless landing possible and then hope that bursts of asymmetric power would serve to keep the aircraft straight. Luckily the wind was slap along the runway-in-use, otherwise I would have had to find another aerodrome where it was so.

The Beau, after a comfortingly accurate three-pointer, ran absolutely straight for a few hundred yards. Then, in spite of full rudder and instinctive jabs with the thumb on the useless brake lever, it began to swing to starboard. Full power from the starboard Hercules failed to check it and the thing left the runway in a rapidly tightening turn, during which it slid bodily sideways on the muddy grass and ended up travelling backwards at quite a smart clip, with the tailwheel protesting noisily. The hidden blessing was that the aft cg no doubt helped to prevent it from completing a real ground loop–as opposed to that normally so described. Thanks to the Beau's solidity, the only damage done was that to the *amour-propre* of the unit's chief test pilot.

The early Beaufighters, particularly the Merlin-engined Mark II, were somewhat uncomfortably unstable fore and aft, particularly during the climb. The fault was relatively mild and was cured by fitting a re-designed tailplane with 12 degrees of dihedral. This was tested first on a Mark II, became standard for later marks and was installed retrospectively to some of the Is and IIs. The dihedral tail probably made the Beau almost too stable for night fighting operations, so some squadrons left their aircraft unmodified.

The complication of the Beaufighter's cockpit was an initial impression rather than a fact. Things were placed where they could be put conveniently, rather than where they should have been placed logically. So, apart from that splendid and universally fitted invention, the standard RAF instrument-flying panel, the single-seat cockpit was a fairly untidy place. However, we preferred it that way, and even liked the multifarious and independently conceived controls which were so typical of British aircraft at that time. The electro-mechanical undercarriage indicators of the Beaufighter, for instance, seemed to have been designed after a look at the drums of a fruit machine, or one-armed bandit; the engine and propeller controls were in two quite separately constructed gates, with the blower and other engine controls in a flat, horizontal panel behind them; and the fore-and-aft trimming wheel was mounted on its own little pedestal to the right of the pilot.

If getting into a Beaufighter's cockpit was quite an involved and athletic process, getting out in hot blood must have been even more so; luckily I never had to try. Entry was made through a belly opening, behind and below the cockpit, via a balanced hatch-cum-ladder. To get into the pilot's seat (where your parachute had been placed already) the hinged seat-back was lowered; you then climbed over this, trying not to move any of the controls or kick the instrument faces whilst doing so.

For an airborne exit the drill was a reverse of the process. After jettisoning the torpedo (if you carried one), undoing your harness, lowering the seat-back and triggering open the trapdoor (which was swung by air pressure into the vertical position, where it locked and formed a windshield), you pulled yourself back on to a pair of parallel bars and swung on these like any gymnast about to do something clever. In this case the cleverness was that needed to drop cleanly through the gaping hole below without ripping your parachute in the process. Goodness knows how one managed if it there were heavy "g" loads, or if the aircraft was, as well it might be, in some outrageous attitude – such as upside-down.

But the Beaufighter was a satisfying aeroplane as well as an effectively functional engine of war which, in its developed Mark X version, carried, without complaint, everything but the kitchen sink. It would, no doubt, have carried that, also, if required to do so for some operational purpose. □

CHAPTER TWENTY-FOUR

Spitfire Score

*

SPITFIRE

TO THE ORDINARY PERSON, then as now, the word "Spitfire" meant just one type of fighting aeroplane – if a rather special one. For the test and ferry pilots the word conjured up memories of a series of aeroplanes, each of which was, to a greater or lesser extent, unlike the others. Through more than a score of different variants the Spitfire had, by 1944, grown to twice the aeroplane it was when the name had been given to it in the middle 1930s. From the gentle Rolls-Royce Merlin-engined Mk I, II, V, VIII, IX and XI it progressed to the Griffon-engined XII and the even more purposeful XIV variants and the very different Mk 21 – when the somewhat clumsy Roman numerals were dropped. Yet all of them, in their different ways, retained, at least until the Mark 21, their essential "Spitfirishness" once they had been manhandled, sometimes necessarily forcefully, into the air.

There were oddities, too, among the earlier variants. The Mk I, even after it had been fitted with a two-pitch propeller, was something of a handful for the amateur ferry pilots of the very early World War Two period. The undercarriage had to be raised and lowered by a hand pump on the right, so the "stick hand" had to be changed after take-off and, with the Spitfire's very light elevator control, one tended to climb away with a see-saw motion as the left arm, then holding the control column, followed, in reverse, the action of the right arm which was doing the pumping. The less said about the Mk VI and Mk VII the better. These were early high-level photographic reconnaissance variants; in the Mk VII the pilot was, as we then put it, "nailed into his coffin". In other words, the air-tight canopy had to be fitted after the pilot was in his seat – though the canopy was ejectable in emergency.

The Mark XII was the first of the really "different" Spitfires, powered as it was by the Rolls-Royce Griffon engine in place of the Merlin. So I shall begin with the story of troublesome experience with this variant.

□□□□□

The behaviour of EN225 had ceased to be anything much better than a joke in rather poor taste. No aeroplane when it first came out of an RAF maintenance unit's workshops for air test after re-equipment, modification, or overhaul could be expected to be in perfect shape. The lateral trim, for a start, would certainly

require attention and adjustment might be needed for half-a-dozen items, including the propeller's constant-speed unit.

But this Spitfire XII had some very new kind of gremlin aboard. It had been brought back from its first post-modification test with a report of erratic variations of power and engine revolutions. In succession, after this and later tests, the constant-speed unit and even the propeller had been changed and various major replacements had been made to the controls and carburation system of the Griffon engine. None of this work had made the slightest difference. The power and revolutions had continued to surge, sometimes with considerable violence, and not a sign of this trouble could be reproduced on repeated ground runs of the engine.

So–with the maintenance unit's engineer officers and the CO himself becoming increasingly suspicious and impatient about these repeatedly adverse test reports–I decided that, as the senior test pilot, I had better take this Spitfire over for a private investigation and to give it, so far as I was able, "the works". I planned a series of progressively faster runs at different heights; these were designed to pin down the airspeed/boost/rpm circumstances of the power surgings–which, I was to discover later, amounted, at high speeds, to dead cuts of a quite frightening violence.

I noted down the effects of different boost pressures and engine-revolution settings as speeds were progressively increased. It did not take long to discover that the airspeed was the critical factor. Power surging would start at about 250 mph with varying intensity according to throttle and engine-speed settings, but below about 300mph the effects were not particularly terrifying. Above a speed of 350mph things really began to happen and a final dive to about 400mph produced a series of power alternations which were so violent that I thought the Griffon was going to leap out of its engine bearers.

During the peace and comparative quiet of a throttled-back descent into the circuit I looked at the assorted figures on my knee-pad and the light began to dawn. Something must be happening to the air-intake system which caused such big variations in ram-air pressure that the automatic boost and propeller controls just could not cope with them.

Other aircraft were waiting for test and I never tried to find out whether and why the intake had been made of so thin a metal that it distorted under air pressure. But a replacement intake assembly certainly cured the trouble completely; Spitfire XII EN225 was cleared for collection on its next test flight.

It was not entirely the extra power which made the Griffon-engined variants of the Spitfire so unlike their Merlin-powered predecessors. Because the propeller turned the opposite way these marks also swung the "wrong way" on take-off. The Merlin Spitfires tried, fairly gently, to swing to port, but the Griffon versions tried (very much less gently) to swing to starboard and full port trim as well as hard rudder pressure had to be applied to keep them straight.

With all that extra power, too, things really began to happen on take-off if you applied too much boost too quickly at the start of the run. Even with the quietest opening of the throttle, up to no more than 9lb of boost, you had an unusually good view of the runway ahead because the Spitfire wasn't going where its nose was pointing. Luckily, with so much noise just ahead of you (and the Griffon had a *very* splendid bass cruising voice) you couldn't hear the squeal of the tyres which held the balance between heading and track as you peered past the long engine cowling while pressing hard left rudder and holding full aileron the other way in an effort to prevent the port wing from digging in.

Although it was powered by the Griffon the Mark XII did not look, or even feel, so very unlike the Mark V from which it had been evolved. True, it swung the "wrong way" and it needed reasonably firm handling on take-off–but it was the Mark XIV and its successors which looked, and occasionally felt, like monsters. We were, for instance, quite unnecessarily worried, before our first flight in this so-stretched Spitfire, by the size of its five-blade propeller. Surely, we thought, the tips must hit the ground unless great care was taken in lifting the tail. In fact, the Mark XIV left the ground so soon, even when using limited power, that you didn't have much time to do more than keep the long nose aimed more or less in the right direction.

Monsters though they looked, most of the later, big-engined Spitfire variants still retained essential Spitfire characteristics. The rudder, because it had a servo balance tab, felt different and you could, if heavy footed, yaw the XIV about rather wildly when diving it on test for the first time, or when making the run-in during a cannon-firing outing–but other controls were normal enough. As with earlier marks, you could, if you liked it that way, bring it in on the approach in that slow, nose-high, lightly powered turn which gave you a continuous view of the runway until you finally straightened out for the hold-off. Only with the Spitfire 21 did the familiar characteristics begin finally to disappear. This variant –which should, perhaps, have been given a new name–was nice enough, but it was not really a Spitfire any longer. It had, for instance, servo tabs on the ailerons as well as on the rudder and a slightly different undercarriage geometry.

For the pilot, three quite minor features typified, more than any others, the Griffon Spitfires. These were the enormous, long-travel throttle lever with its knurled gunsight-adjustment grip; the interconnected throttle and constant-speed propeller control; and the Coffman combustion-starting system.

The giant throttle was probably necessary with so much power available and with the need to limit this power accurately on take-off. But, though one quickly got used to it, this big control seemed to be clumsily out of keeping with the still-dapper characteristics of the Spitfire. The interconnected engine and propeller controls were probably accepted as a trouble-saving device by the operational pilots and forgotten–just as the automatic boost control, that so-English invention, had been accepted and forgotten. But for the test pilots and their technician ground crews these controls are remembered with a mixture of irritation and pleasure of achievement.

Compromise adjustments of the throttle and propeller-control rods, the boost-pressure control capsule and the throttle-gate stops had to be made so as to produce results which gave approximately the right engine revolutions to match the boost pressures throughout the throttle range. This compromise involved a series of interrelated adjustments which, for various reasons, could not be made successfully with the help of engine ground runs. The fitter had to work on a series of apparently disassociated engine-speed and boost pressure figures which were brought back by the test pilot on his knee-pad. With practice, the fitter nearly always got it right first time, so that on the second, or, at the worst, the third flight, the revolutions would match the boosts, as near as made no matter, throughout the throttle range.

The combustion-starter system–whereby something resembling an enormous shot-gun cartridge was fired to produce gas pressures sufficient to provide auxiliary power capable of whizzing the engine round–was fine so long as everything had been set correctly for a clean start in the particular conditions. Otherwise it could cost HMG a lot of money in wasted (and very expensive) cartridges. With electric starting and a power trolley you could play games with the throttle control and the priming pump while turning the engine round until it chose to fire. Not so with the combustion starter. This gave you only one start per cartridge; so everything had to be dead right, with the correct throttle setting and number of strokes of the priming pump.

When on ferrying trips with a Griffon-engined Spitfire, or when paying technical visits to other units, you put a couple of cartridges into your pocket and hoped that these, and the cartridges in the starter "magazine", would be enough,

or that the visited unit would have a stock. Without cartridges your Spitfire could be just about as mobile as the Rock of Gibraltar.

On one test flight with a Mark XIV I got myself lost after a series of above-cloud speed checks and landed (at Lasham, as it turned out) to ask where I was. With uncertainty about the magazine-supply status and with no cartridges in my pocket, I had to break all the rules and leave the engine running with the brakes on while I got out and ran to the controller's chequered box to ask the humiliating question.

□□□□□

But I really began to learn about the Spitfire and about test flying during a period in which I was seconded from the RAF to help with production-line testing of the Marks VIII and XI. Vickers-Supermarine was the only factory/assembly plant to make the VIII–which paradoxically, was a refined version of the mass-produced IX. The XI, equally delightful to fly, was the last photographic recon-

LEFT: *This illustration shows the cockpit of a Spitfire which Vickers-Armstrongs restored to an airworthy condition. This aircraft has the external finish of the Marks used in the Battle of Britain; the controls and instruments are more representative of later variations, such as the Mk VIII* (ABOVE).

naissance variant with the Merlin engine. At that time, in the early autumn of 1944, the company's experimental and test development base was at High Post, near Salisbury, Wiltshire. From here the production-test pilots commuted themselves by air, using a variety of "impressed" civil aircraft, to and from the dispersed assembly plants at Chattis Hill (near Stockbridge), RAF Keevil (near Trowbridge) and RAF Aldermaston (near Newbury–and famous later for the protest marches on the nuclear plant built there).

A lot of things had to be remembered or re-learned during a second experience, after a two-year gap, of production test-flying.

As the Spitfire VIII bored upwards towards the overcast sky, I checked once more around the cockpit and made the very conscious effort of tearing my mind away from the outside world and concentrating on the blind-flying panel. It was five years since I'd done any really serious instrument flying. Now it had to be done well and accurately. The facts and figures recorded on my knee-pad would be checked afterwards by the chief production test pilot. So they had to be good. It was just unfortunate that this first day's weather suffered from ten-tenths cloud (as we said it then) at 3,000ft.

During four years or so I'd been testing aeroplanes at RAF maintenance units, but, even as the Spitfire had been settled carefully into its 170mph rated-power climb, the practical differences between the two kinds of work were already beginning to be obvious. At the Service units aeroplanes were checked in a general way, using routines which had been modified by growing individual experience; things worked or they didn't work, and it was left to us to decide whether the handling in various arbitrarily chosen conditions was either reasonable or not.

Now I had to bring back a series of acceptable and useful comments about a straight-off-the-production-line aeroplane and a set of really accurate performance figures which would qualify it, on this and later tests, as being up to acceptable standards. Hence–the weather being what it was–the over-anxious concentration on the six instruments which made up the standard RAF blind-flying panel of those days.

As it turned out there was no particular cause for anxiety. Instrument flying is, like bicycling and swimming, a trick which, once properly learned, is never entirely forgotten. The simple basis of success is confidence. You learn–very laboriously at first–to ignore all physical sensations and to react only to the information provided by the instruments. And pretty sparse and difficult this information was in the pre-war basic trainers. Fortunately, the instruments in this Spitfire were very much more easily followed than those under the hood of the Tiger Moth in which, five years before, I'd been passed out during an instructor's course. The artificial horizon was just what its name implied and faith was all that was needed to be able to fly on it successfully; the directional gyro was something I'd been regularly using as a heading indicator on my way between the As and Bs of the world of testing and ferrying–so there was nothing unusual about that.

The cockpit of probably the only Spitfire XIV still flying. Used once for development testing of the Griffon engine, it later served as a communications aircraft for Rolls-Royce test pilots.

Nevertheless, as the cockpit darkened, I found it to be absurdly difficult to hold the airspeed needle steady at 170mph, no more and no less. Why? There could be no reason, I told myself, except over-anxiety–and forthwith loosened the insane grip with which my hand, all by itself, had been holding the control column. I learned again to relax and to let an aeroplane do what it can usually do so much better by itself.

But, when the gloom showed no sign of lessening, I began to suffer from nerves. I hoped that it wouldn't be necessary, on this very first trial test, to do the approved speed check while flying on instruments. This involved a sustained level run at exactly 10,000ft and I knew that I couldn't do this well enough in cloud for my test-pad figures to be worth much more than the card on which they were written.

All was well. As the smaller of the two altimeter needles passed slowly through the figure "8" for "8,000" the cockpit began to brighten and a few seconds later I was bathed in hot sunlight.

But I could not yet be freed from those hypnotic instruments. The point in holding rated power at an exact climbing speed was the important one of catching the engine's boost needle just as it started to fall away and noting the altitude; "full-throttle height" it was called. Only when this little chore had been done could I relax. The figure noted, I looked up. Except for a few streaks of cirrus the sky was clear. Below, the unbroken cloud layer was already moving more and more slowly as the Spitfire continued to climb up to the altitude at which the supercharger would (I hoped) automatically alter its gear ratio.

Although I was waiting for it, the "kick" from the Merlin engine, as the blower changed gear, came as a frighteningly unexpected interruption of normality. But I remembered to write '13,200ft" on the test pad before starting to descend to the appointed level-speed-check altitude–3,000ft or so below me, and, fortunately, just about 500ft above the tops of the cloud layer.

One feels dangerously safe when flying above cloud and it is always difficult to force one's mind into the sensible state which really accepts the fact that, down below, there may be gloom and rain and some little difficulty about finding one's way back home.

The chief test pilot had explained the team's simple little rule-of-thumb method by which (if too many repetition tests did not confuse the result) we could hope to finish up within a few miles of base. The rated climb was made up-wind; the speed run was made on a reciprocal heading downwind; and the concluding terminal dive was made up-wind. Theoretically, the aeroplane should arrive back more or less over the take-off point.

I hoped the trick would work for me. If not, I intended to fly south to the coast. There I might reasonably expect to be able to recognise some feature which could be used as the starting point for a gentle map-reading saunter back to our main base at High Post and, thence, to the Chattis Hill dispersal-factory field from which I had taken off on this test.

The Spitfire was trimmed carefully into its level-speed run and I thought how fortunate I was to be testing a Mark VIII–almost the ultimate natural development of the variants which I had first flown in 1939–40. There had already been a score or more of different Spitfire marks in service and we had sampled nearly all of them at the maintenance unit from which I had been seconded for factory production test-flying.

With the level speed checked and the slightly better than standard figure "332" written on the knee-pad, the Spitfire was turned through 180 degrees and nosed down towards the cloud layer for the final power drive. It was a nuisance that it had to be completed in cloud, but at least this would force a concentration on the behaviour of the aeroplane as such. There would be no opportunities for wool-gathering in admiration of the landscape.

The Spitfire plunged into the layer–the pattern of apparently hard shapes dissolving instantaneously into swirling, amorphous grey vapour–and I settled down to the business of pretending that the two-by-four world of its cockpit was my entire universe. It was not advisable, or even possible, to think of things outside the bullet-proof screen and the bubble canopy, or of events other than those which were being measured by the needles of the instruments and the feel of the controls.

As I started to ease out of the dive the real world of Wiltshire's downs, little spinneys and grey villages was already waiting there beyond the flat-topped engine cowling and I wondered, watching for something recognisable to come up, whether the rule-of-thumb navigation system would work for me.

It did. There, ahead, was the long, unmistakable scar of Boscombe Down's new runway, with the grass aerodrome of High Post beyond it and the spire of Salisbury Cathedral rising from the valley on the left. I pulled round in a tight turn towards Chattis Hill, which I knew now to be only a few miles away to the east.

When I had landed up-hill on that little grass field and had started to taxi in I could already see the chief production test pilot waiting there on the tarmac to take the Spitfire once again through the same routine and to check my facts and figures. It was impossible to believe that I'd been airborne for no more than half an hour. □

LA2
2
JX

Epilogue

Over the public-address system the captain of the Stratocruiser had told us that a non-stop New York–London flight was planned. This was mildly unusual even for an eastbound North Atlantic service in the mid-1950s. Crews usually played safe with a refuelling stop at Gander, Newfoundland, eastbound, and at Shannon, too, on most westbound flights.

In the event, we got no farther than 'overhead Nova Scotia'. Soon after night had fallen and the Stratocruiser had settled into its second-step cruise, I noticed, sitting forward on the port side, that the cowling of the inboard engine was occasionally giving a violent twitch.

The flight engineer came back to peer at it. After a few more cowling movements – more violent and frequent and now accompanied by faintly heard pops and bangs – its propeller was feathered and the captain told us, in due course, that we were returning to New York.

Even hardened passengers become just a little more talkative when this kind of thing happens. At New York International I had met, almost for the first time in a decade, a one-time test pilot acquaintance. We were sitting together and when the propeller was feathered we made the inevitable ham jokes about the statistical chances of multiple engine failure and about the Stratocruiser's ability to fly on two engines – or even, maybe, on one.

It was very dark outside and the Stratocruiser had a long way to go at its much reduced three-engine cruising speed. We started to reminisce about the characteristics of wartime aeroplanes which we had flown. He had been a real test pilot – not, like me, a checker of production-line aeroplanes or of maintenance-unit modifications – and there were a lot of things I wanted to know about accidents and incidents which had never been properly explained through the usually reliable Service bush telegraph.

He knew about most of them. Presently we got around to discussing the Blackburn Botha – that never-to-be torpedo bomber which went through fatal periods of trouble on production and other test-flying work.

The first Botha I saw, in the summer of 1940, was not recognisable as such. I had taken a taxi Anson up to Abbotsinch, Glasgow, to collect a Botha and to deliver it to the maintenance unit at Kirkbride, Cumberland. During the landing circuit I noticed a trail of aircraft wreckage crossing two fields.

The accident had only just happened and the test crew had been killed. So I asked no more questions – I just went over the Botha's cockpit arrangements with one of the flight-shed engineers and took it off on its delivery flight.

Later, as a maintenance unit test pilot, I was to find that the Botha was a handleable and well-laid-out, if operationally useless, aeroplane.

As we understood the situation, the Botha had had to manage with Bristol

WARNING
LEFT ← → RIGHT
BOOST
PRESS
BOOST
PRESS
ENGINE SPEED
HUNDREDS of R.P.M.
AIR
KNOTS
ARTIFICIAL HORIZON
CLIMB
FT/MIN
DIRECTION INDICATOR
L
R
COLD
AIR
OXYGEN REGULATOR
DELIVERY
NORMAL
FIRING
SWITCH
OFF
STBD

Perseus engines, while the limited production of the more powerful Taurus powerplants from the same factory had been allocated (not unreasonably) to the Bristol Beaufort torpedo bomber. So we had our little joke–that the Botha could, *of course*, carry a torpedo, but only with near-empty fuel tanks, and that it could, *of course*, go an awful long way if it was not expected to carry anything.

So it sat sadly about in MU dispersal fields and satellites and was used seriously only by training units.

The MU test pilots knew it as an easy-to-fly device which was shuttled between the units, where it was maintained, modified and equipped, and the dispersal fields in pleasantly out-of-the-way places like Anglesey.

All the same, the Botha worried us more than a little. We knew about the accidents, but the bush telegraph had failed to tell us why these had happened, and the powers-that-be had also failed to tell us (as they usually did).

We, too, had our own little bothers. There was a batch of Bothas on which, we discovered painfully, the engines would cut dead if take-off power were held for too long. It turned out that the stops of the main fuel cocks did not quite register; when these were turned to the "open" position the valves were not fully open. Fuel starvation could then build up to a dead-cut situation.

My talk with the test pilot–15 years later in the Stratocruiser limping back to New York–demonstrated the desperate knife-edge on which aviation events are so often poised.

The cause of the earlier Botha accidents was, in itself, simple enough. In those days lateral out-of-trim-ness was often adjusted by the simple and effective means of doping lengths of cord on the upper trailing-edge of one or other of the ailerons. If, for instance, a test pilot reported his Hurricane to be "left-wing low", the cord was doped on the left aileron. This caused the aileron to be pressed down by the airflow and to lift the wing. With the Spitfire the same effect was obtained by "planishing" the metal trailing edge of the aileron to the shape required.

We became very clever, with experience of different aeroplanes, in judging the amount of cord required. Instead of saying that "this aeroplane is a little left-wing low" we would say that it was "four inches left-wing low". Four inches of cord would then usually come near to doing the trick.

For some now-forgotten reason the flight-shed crews for the Botha adjusted the trim by gluing cord or battens *under* the aileron's trailing-edge. This could cause the aileron to ride up to such an extent that something had to give.

The cause of the accidents might have been discovered more easily if the eventual investigation had not been so thorough and basic. A test pilot and team from Farnborough sensibly decided to start from first principles. They took Bothas straight off the production line and went right through the test routine. But they corded the ailerons on the upper trailing-edge–so the circumstances leading to the accidents were not repeated. Another pilot was killed before the cause could be pin-pointed.

Mistakes, I thought, as I sat in that Stratocruiser, become translated into hoodoos only when the lawless laws of probabilities and permutations bring several mistakes together at the same moment in time. The chance that a tossed penny will turn up a head is still equal after a dozen heads have been thrown.

What, I wondered, was the statistical likelihood of mechanical or other failure

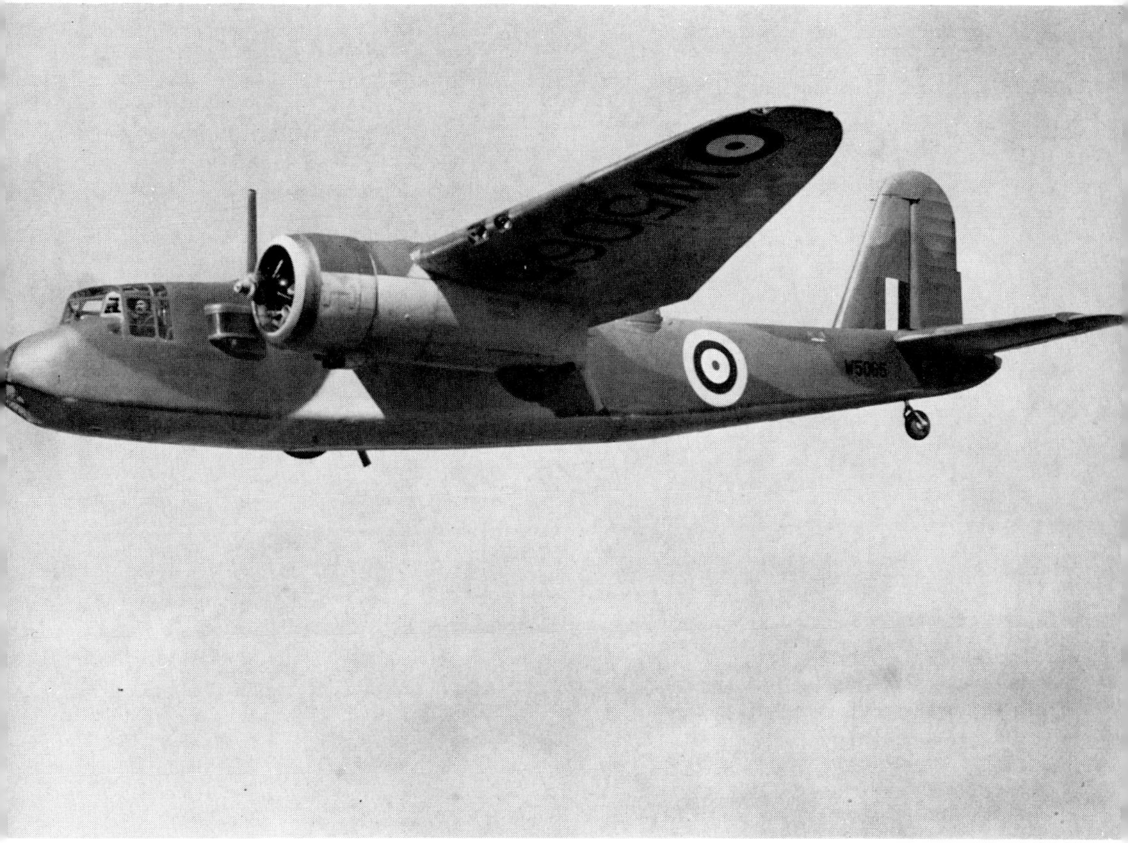

in each one of three haphazardly chosen powerplants during a five-hour period? Obviously it was very, very small.

But two cynical old-timers on that Stratocruiser flight were glad that the bar service remained unaffected during the routine emergency return to New York.□

Acknowledgements

We are grateful to the consultant editor of Air BP *for permission to borrow the following blocks*(*page* 2, 9, 14 (*right*), 18, 19, 25 (*upper*), 26, 37, 38, 59, 61, 67 (*lower*), 69, 72 (*top*), 73, 78, 86, 99 (*lower*), 106, 110, 111 (*top, centre and foot right*), 116, 118, 121, 130, 131, 135, 142 *and* 144.

The original source of photographs, where known, is given below.
The Aeroplane: 14 (*centre right*), 15, 33, 38 (*lower right*), 53, 56–7, 61, 72 (*lower*), 86, 88–9, 93, 96 (*lower*), 111 (*foot*), 116–17
Flight International: 20, 25 (*lower*), 31, 35, 38 (*top and lower left*), 40, 47 (*lower*), 66, 72 (*upper*), 73 (*upper*), 92, 95, 99 (*lower*), 103, 104–5, 107 (*centre*), 121 (*centre left*), 123
Ministry of Technology (Crown Copyright Reserved): 14 (*centre left*)
Imperial War Museum: 37
Cyril Peckham: 47 (*upper*)
Central Press Photos: 51
Air Ministry (Crown Copyright Reserved): 55, 85
Associated Press: 67 (*upper*)
RAF: 73 (*lower*)
US Navy via Naval Aviation News: 110